1. Introduction

OK! Welcome to Polis. I'm Abon, designer from Aldebaran Robotics. And I'm doing this training next three days. Basically my expertise is not in MTI, OK. It's mean I'm not MTI. My expertise is more about robots here. Basically this training just show you how program you're robot. Then because you need to play back you're entities. I have some deepens key's here and there key's you can copy to use this material, software, and some document you can copy on your computer. Maybe you don't now, license is link you're Choregraphe, and there is the simulator's. I want start training by introduction hardware this robot.

2. Hardware description of the robot.

SO, this morning I want introduce you hardware and all so show you this morning more software. And this morning I'm will all so show you nao picture. This afternoon I will more excuse about Choregraphe witch is more program mobile. And then to morrow i will introduce you some progress programming. And then afternoon I'm can answerthe question some nesses ice. Last days we have workshop, try to practice, and answers your questions.

2.1. Characteristics of the robot

SO, let's now find the sensors and actuator robots, and then I'm show best practices. Power for you maybe is not critical because you have one robot. Remaining to select and this best practice is just employing you to robot. So, falling sensors and actuator have some join, led, one

00:05:00

Inertial board, 4*2 FRS, some switches some button, cable some, and we have infrared emitter/ receiver, 2 cameras, and 4 microphones, 2 loudspeakers. Almost this sensor's you can make every think.

2.2. The switches.

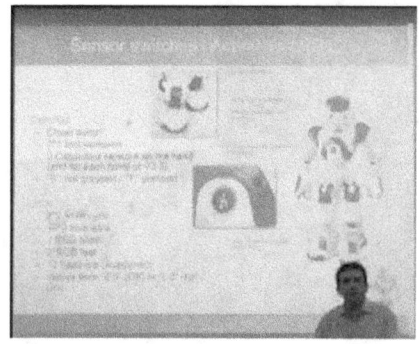

So let's start with the easiest one, switches and capacitive. So you have the chest button here. This button is used to switch on the robot and switch off the robot and also to get its IP address. You also have a foot bumper here. So there are 2 contactors per foot for a total of 4. We have 3 capacitive areas on top of the head and we also have 3 capacitive areas on each arm. So the switches and capacitive sensors are binary sending 0's and 1's. For the next we have 8 sections of RGB LEDs for the eyes and 10 blue LEDs on the ears. There is one RGB LED on the chest, the button. The color indicates battery levee but you can override this. We also have one RGB LED prepare and 12 blue led on top of the head. The LEDs also operate on 0's and 1's. So I will demonstrate a switch.

2.3. Monitor

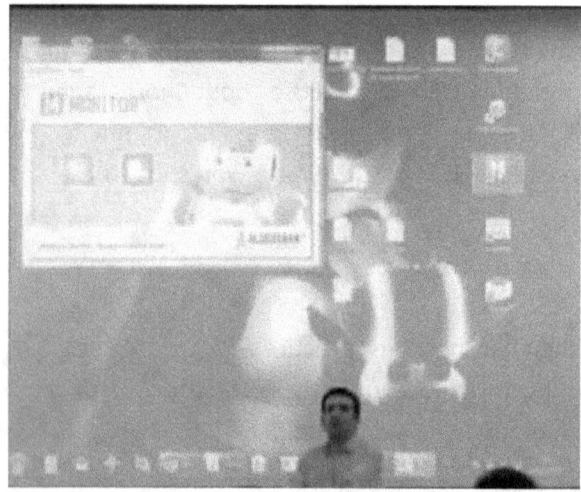

Monitor is a kind of debugging tool; it is simple software to monitor different values from the robot, a sensor, joint position or other values. It's a very simple program and only has a few buttons.

First you connect to the robot; after you connect to the robot you have this window. If the robot is connected through WIFI the router is here and my pc is also connected to the router. So of course, when you want to connect the robot and pc need to be on the same network. So in this window you can create a new configuration file.

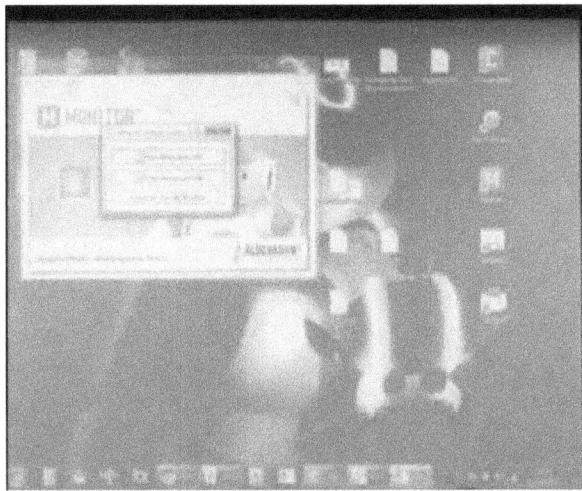

So what is that, when you look there are many buttons,

00:10:00

If you want to look at a few you can choose the ones you want and then you can save this as a configuration file.

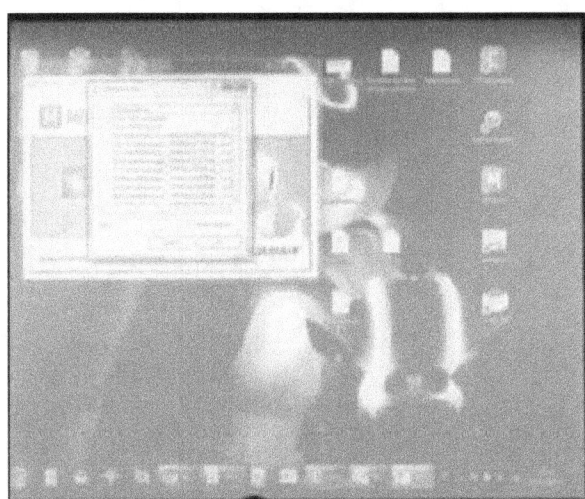

So later we can save configuration file, and chose again. So I'm will show you configuration file.

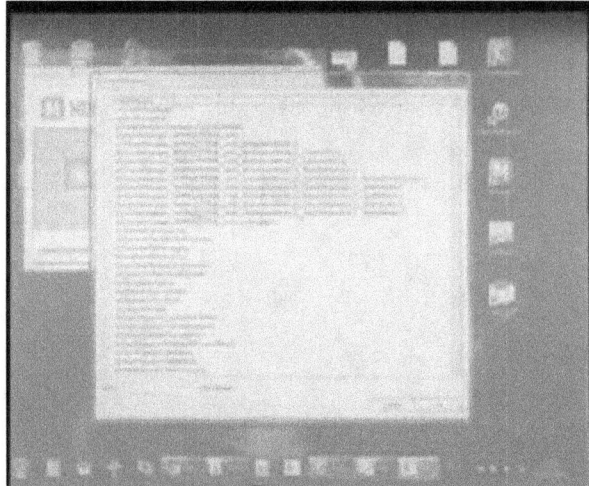

You can see hardware. You can pick you're device and to begin monitor.

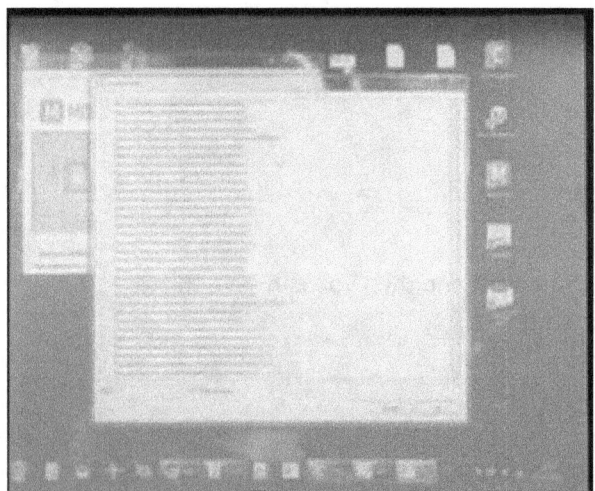

You pick this wan this wan and before used you can save it.

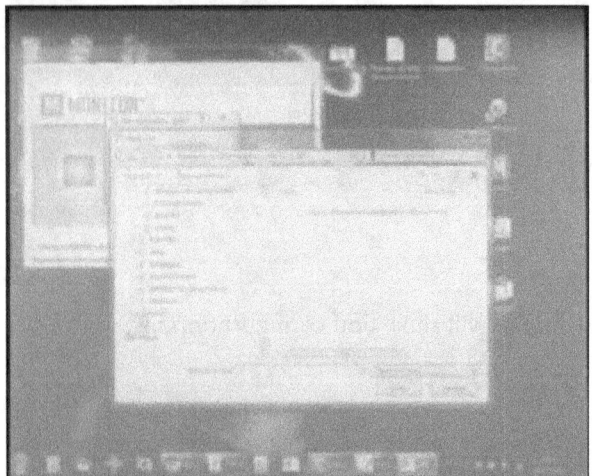

This is folder yours pc, you get to choose them again.

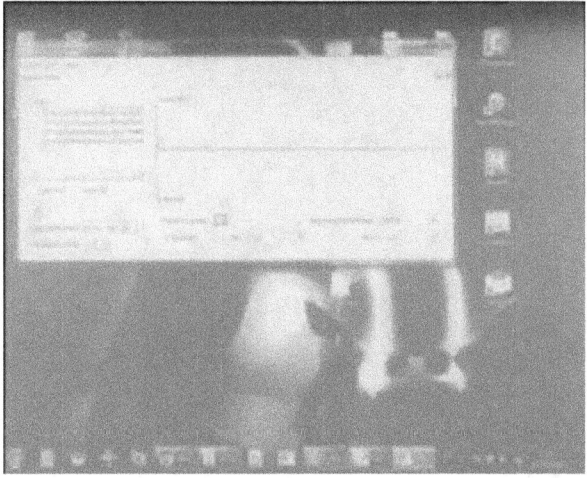

When you do that you already have version file. Connect again then I'm choosing open version file.

So I'm taking choose hardware, and I'm have monitor space implies two window. You have window hardware monitor, and you have program to display time. Here you have volume scale, it can be check it or none check it. Then you locked scale and wild access device specified. If you want to see volume, you should also click the graph here.

Complete Choregraphe volume display here color. And if you check graph it also display here. You can choose graph. This line is fit button I'm can see that. This switch in zero to line and you can see different color.

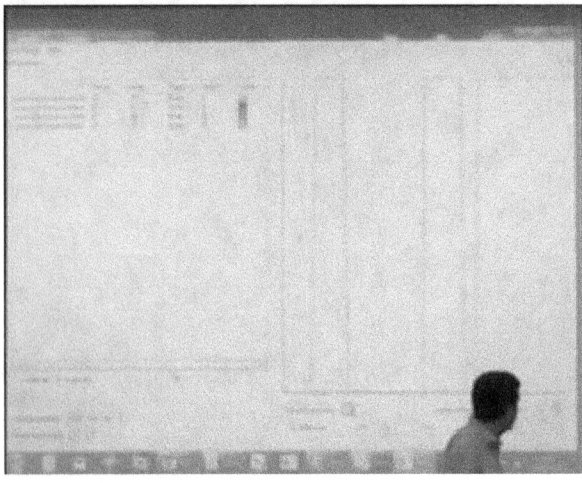

So this line light yellow have two push and you can see. You can change color. If pick color and then push button basically you walking monitor to check that everything is fine with your robot. I'm already showed when click on button chose connect to robot. So you don't have robot, you have simulation, but you can also connect monitor simulator robot. That is you can just take here.

2.4.Main components and actuators

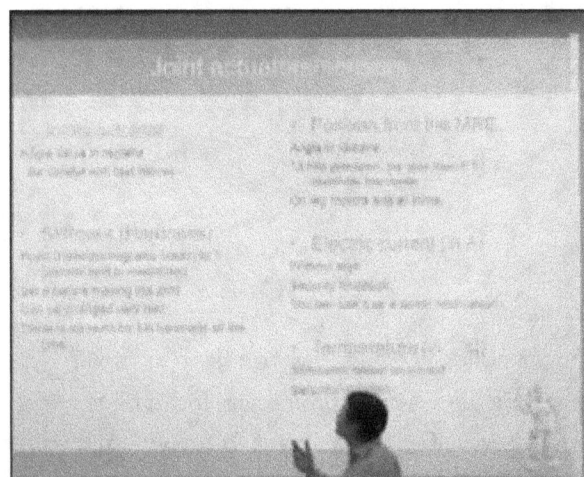

So the next step is the hardware joints, also very important for the robot is the actuator. Nao has 25 motors, 25 degrees of freedom and for every motor you can set or get the values, the angles in radians.

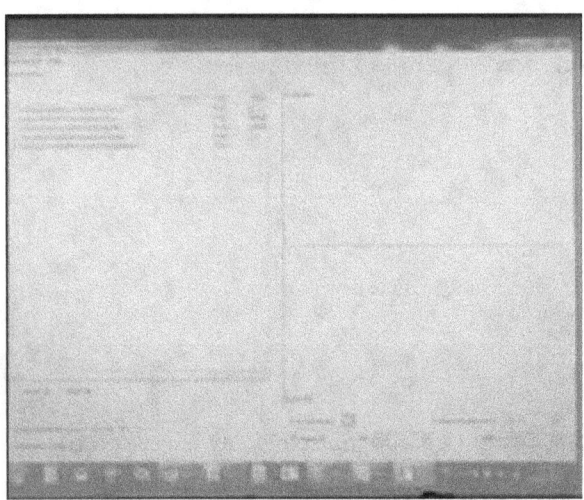

To get the angle we use the MRE which stands for magnetic resistance encoder and it's basically a chip with a magnet on the rotor so when the rotor is turning the chip senses it. It's a very precise system within 1 degree of precision. Also it is an absolute value not related to how many turns. You have 1 MRE for every joint and you have 2 for the bottom joints of the robot, one directly on the motor and one on the gearbox.

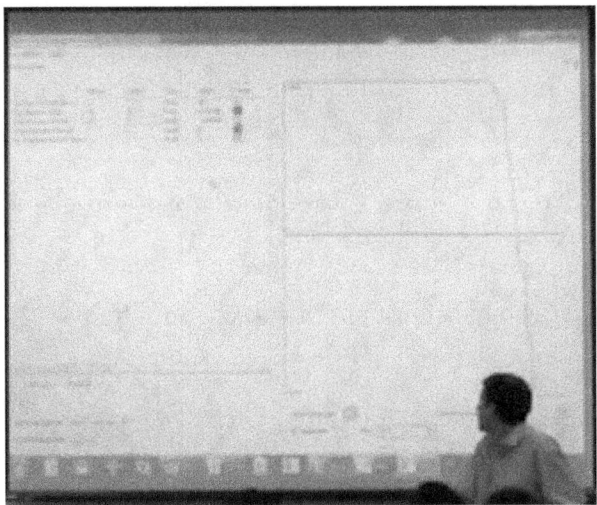

For every motor you can also get electric current and you can also get the temperature in Celsius. The temperature is simulated; there is no sensor on the robot. It is calculated with an algorithm based on the current thru the motor. You can also get and set what we call the stiffness. The stiffness is how strong the joint is. Currently the stiffness is loose and I can move the joints, but if it tries to stand up it will fall. If you set the stiffness stronger the joints will be stronger and resist. So 0 is no stiffness and 1 is most stiffness. We also have some algorithms that will calculate the best stiffness to avoid the motor getting hot too quickly and to help with the battery life. So I'll connect again to the robot and check the right shoulder pitch. First I will wrap the position; I'm basically looking at the data from the right shoulder pitch.

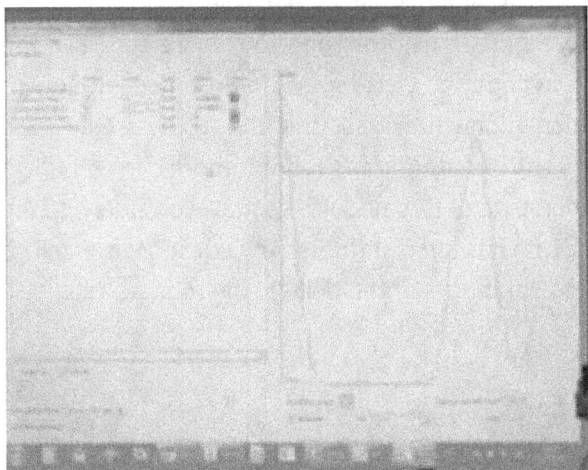

So, yeah, while you move it, you can see the solution at angle is changing in radius. I can also look at the stiffness and occurrence. So and grab this too and i will also watch the temperature of the bottom. But to do that i need to put the stiffness and the radius.

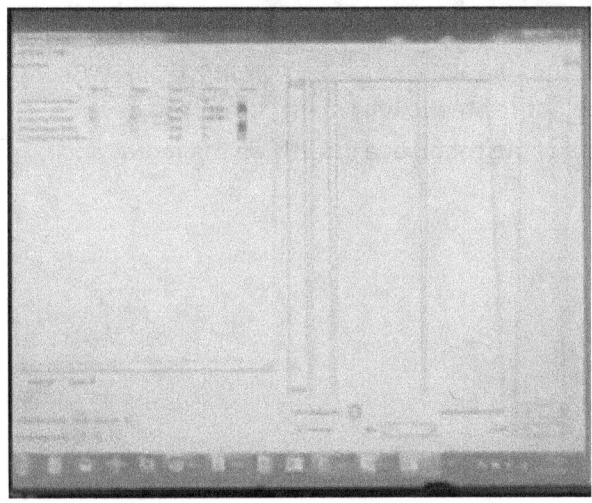

So i would do that by drugging this arm. I will explain this ... So here i just put the maximum stiffness. Like i told you 1 is the maximum stiffness.

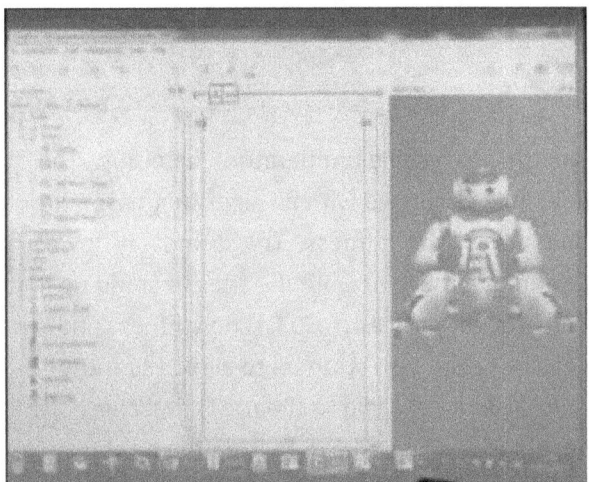

So now we go out a small stiff which moves the robot like that. And in gradient we have armless or stiffness of as the same which is currently around 0.3. So ... And in the other one you have electric current which is quite new. Why? Because the robot doesn't need to sense some, for instance, motion to keep his position. But if I force on the joint, on the articulation you will see that the stiffness will increase. And preferably also increase if you can force it gently on it. And the robot is more currently trying to resist. And if I'll be keeping doing that the temperature of the module is completely nuts? 2.... So yes, this is the stiffness and the current for the joints. That has been stiffness. Any questions so far? next sensor is initial bounce. So we have initial bounce look at it here at the belly of the robots.

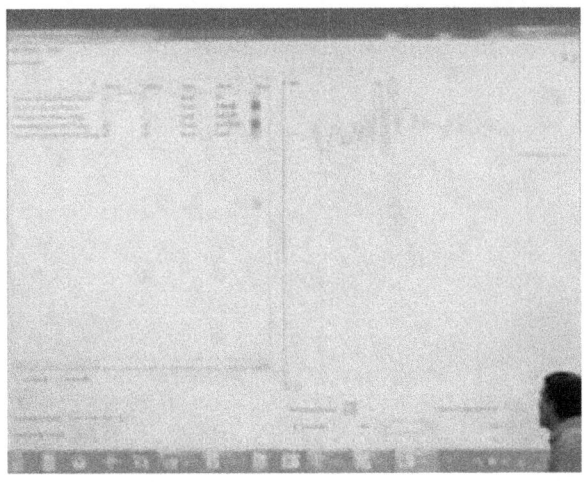

2.5. Sensors

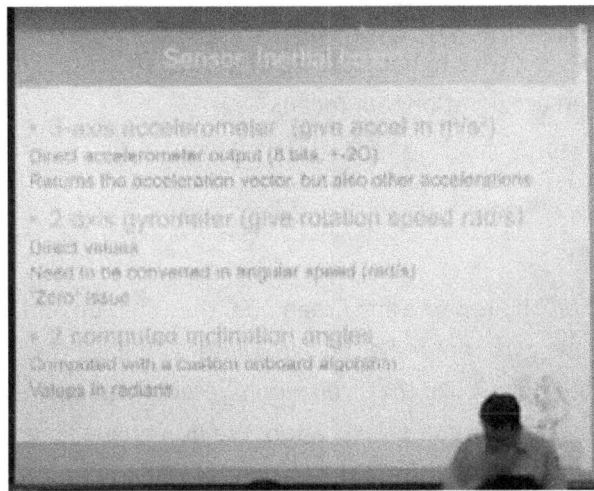

And in this box you have a 3 axes accelerometer and the 2 axes gyro meter. So the 2 axes axel meter gives you acceleration. And the gyro meter give you rotation speed. Also compute all 4 the current inclination of the robot, so the inclination angle of the robot.

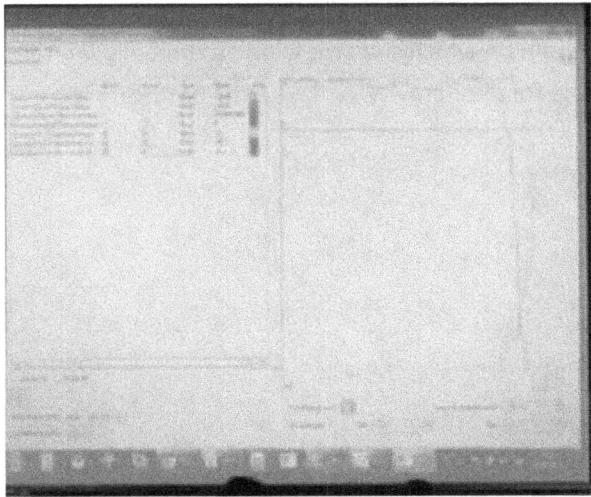

So let's look at that monitor. The first i will graph the accelerometer. Focus on acceleration x, y, z and the axes are like that.

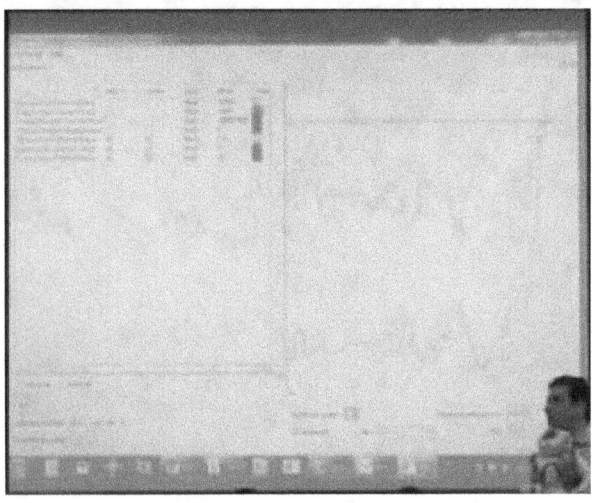

So the x is in front f the robot. Y is on the side and Z - up. So you can test that z accelerometer. X and y almost nil, but z has a negative value, because robot is detecting the force of gravity. Of course if i put the robot like that. And now I'm pressing and now is x blue. X tries to detect force of gravity. If I rotate it like that It goes the red y axes ... Of course i can detect an acceleration that means

00:25:00

Move it like that, like that it will detect tension. Like that. The red one and green one, they step up and down. I can also grub now a gyro meter. Here we grub the gyro meter. So the gyro meter you have the y rotation like that and we also have the x rotation. So if i move it like that, you won't detect rotation. And if I move it like that it's the red code detecting. But we don't have this rotation. We don't have along the z axis. So that's the speed, the rotation speed, which means if i after the rotation is finished the curve is turned to 0. It only detects the rotation at the moment of the speed. And we also have, like i told you, the angle. The angles this time detect the current angle of the robot. When it's like that it won't keep the same angle. And when it's like that we can go back to 0 here. Ok? So there is the difference between the angle and rotation speed. - Does it 0 when it's standing? - Yes, standing is 0, almost. Basically yes. But this is also scaled, so it can be confusing. So basically the sensor ... sensor you can use it detect if the robot has fallen. Like if it's lying down on which side: it's lying on the belly, on the back. So it's like checking the angle you can check also the force of gravity.

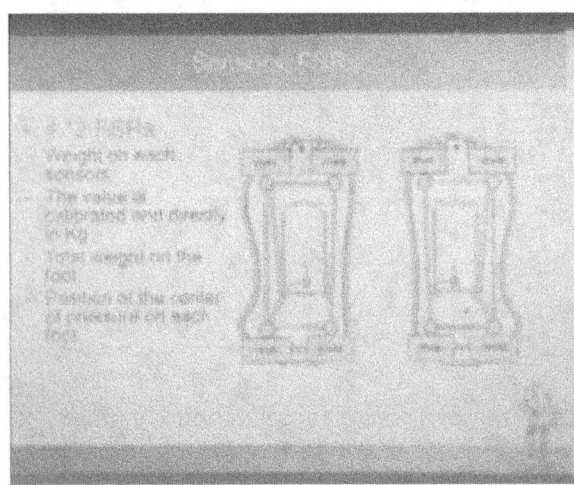

The next sensor is FSR. So FSR means force-sensitive resistor and this is pressure sensor. Ok? So it detects how much pressure it has. Now look at their little foots here, so there are at this little holes here 1, 2, 3 and 4. So you have 4 sensors in foots which means the robot can detect its own weight it's more in front, or more in back or more in side. We also give so called the center of pressure for each foots, depending on its gravity center of the four sensors. We can check where the body distributes its weight - if it's very far from the foot - you may fall. So the value, you can get the direct value from the sensor or you can also get the value in kg. So let's check it on monitor now. Trying to the robot

00:30:00

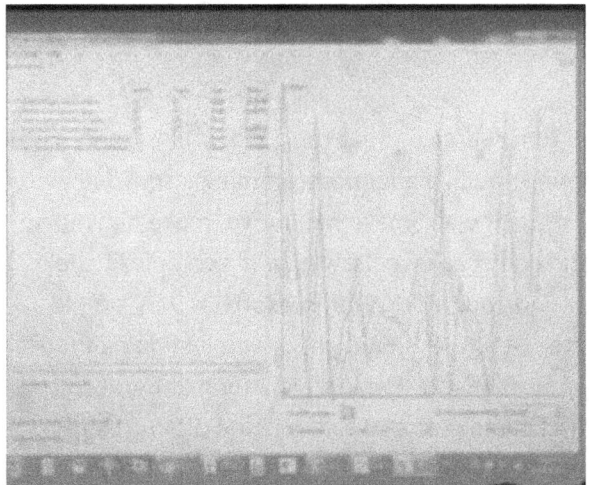

Over graph to the new first of the sensor, first sorry, over right or left foots. Its left foots. So i press this one or this one, this one or this one. Of course if i press somewhere else all 4 values will, all four sensors will have different values. So I can also graph the sensor of pressure.

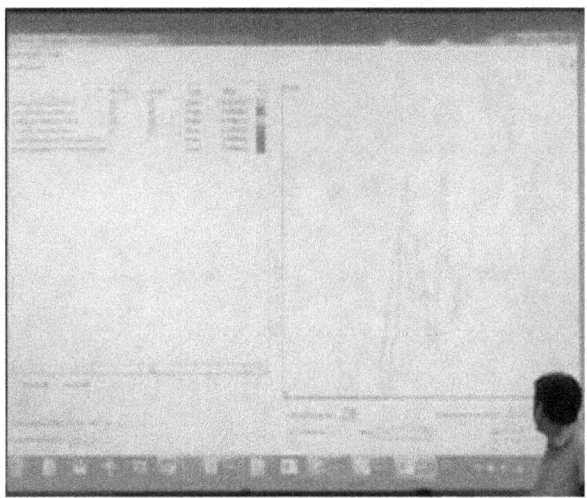

Last leg in the middle, so if that moves up, it would be the x value we may want it to change. You don't want this moves negatively, remember that. And that's possible move left and right to the value which is ... Any questions so far? Yes! -Is there a formula for the rebounding everything to the reaction it in kind of x, y - i think you have to figure out by yourself if you want to make there a close lop at horizon, you have to figure out by yourself.

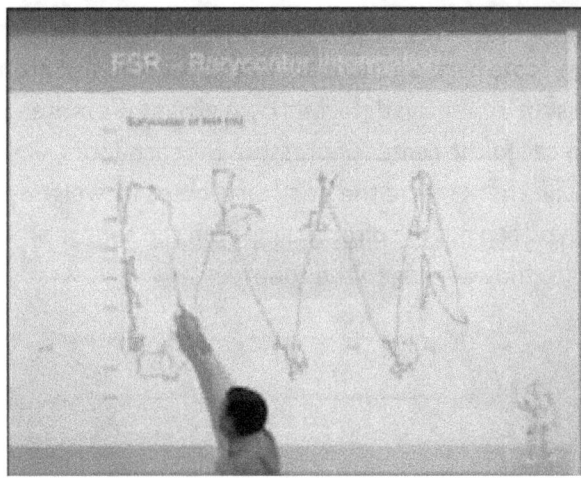

What i would say, for your project i don't think you need to research on that thing. For you is is more application about MTI, I think it's much higher programming, much higher programming also You have researcher that the only want you to {do input various} make it walk, make him move, make him move fast. And of course they will use this kind of {sensorial and control all we draw now it's complexity he'll have to walk here}. For you i think you don't need, we provide our block. That reason it is very simple interface you can. To do display, you will manage each step. And that's enough for you just doing that. But i still show you that, because i don't know may be you can have a simple task, i don't know like checking a pressure. For example a very simple behavior, if someone catches the robot pick up the robot; you see low pressure on foot. Checking that you know the user picked up the robot. And you can make the robot react: "oh, that's unkind, take me home or put me down!" So this kind of sensorial hits back in the head. It really depends how many times you want to spend to develop your robot. So , ah, this is for example on the {Barry} sensor moving while the robot is walking So the robot is walking in that direction and he is moving left and right. That's just an example. it is quite old data actually. The next sensor is the sonar.

2.6. Sonar

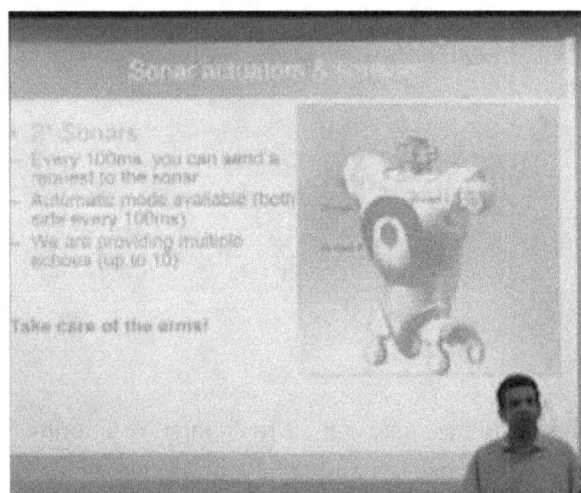

So the sonar we get it on the chest. So here on top is the emitter. On both sides there is a receiver. So there are 2 emitters and 2 receivers So one for the right side and one for the left side. So you can get ... from echoes and pretend {everyone is shaken} and know when you use the sonar take care of the arms, because if you put arm like that in front of the robot the sonar signal will be lost. So you should put arms along the body to use the sonar. The sonar is not stopping automatically when the robot foots for 2 reasons, 1) to save the battery, because we love people use sonar and also because it is making some inter franking noise. So if you want to {reserve} you just need to stop it from API. -What time you should stop it to manage best practice? So may be i can try to demonstrate the sonar.

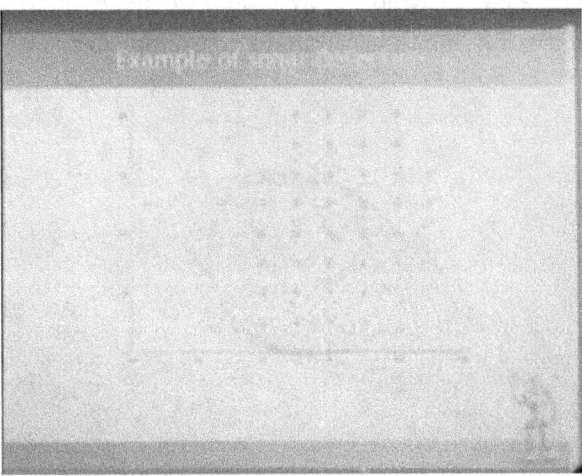

The sonar is usually a very noisy sensor. You should know that. This is just kind of imitation of detector here. It is not very easy to see something, but I will try. So we provide. For left and right antenna, for left and right, but I will just graph 2 of them, that's enough and like i told you i need to stop them. I can show you this graph here.

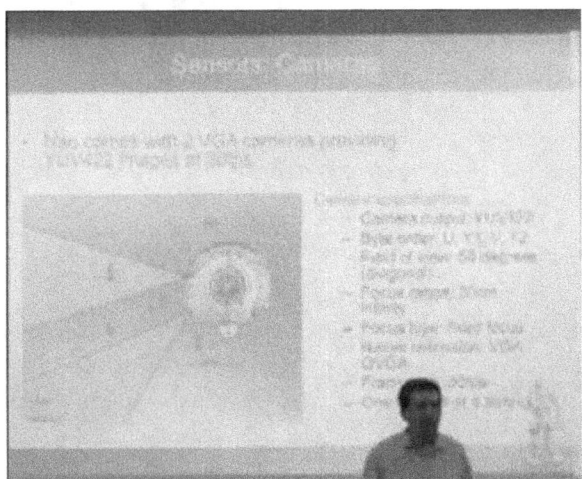

Robots standing, so it gives you normally a distance in meters and if I move it, ok the distance is increasing or after you have the 2 scans it is easy to see that user is here. So at the sonar it can walk between 25 cm to 2 meters maximum 2.5 meters. So here we are around 1.5 meter, i don't have a ruler to measure correct. And we have to move that can increase a bit more to thing. Move it back. I will go to the open 25 which is the minimal. Even I'm closer actually Like I'm at 10 cm from the wall, but it's still hears 25 cm. We can also, if i rotate the robot a bit like that than you, i didn't rotate

00:40:00

it much, then you can see that red curve, red curve is the nest is about the green one, because of course the left distance is bigger than at the right. And if i rotate, it will cross here and then move one which is bigger. So the sonar is working quite well from the wall for example or flat space, because the wave can bounce easily and nicely off the wall. If you have very strange object like that you will have different echo, so you'll have lots of noises. Usually if you use the sonar it is better to use filter to filter up the data and leave it on all the time, something like that. Otherwise it is working quite well if it's a flat wall. For example this one has a slope, so i guess may be will bounce, may be also well. So it really depends on the environment. But this can be useful to make a very simple navigation and to avoid robot bumping to the wall. So basically if you have, if you can detect the distance on the right is closer to the distance on the left, then we can make robot to move to the left to avoid the walls, and we just need to walk it away and save. That's why it's good to have 2 sonars, left and right. So that's an example of different sonar efficiency.

2.7. Camera

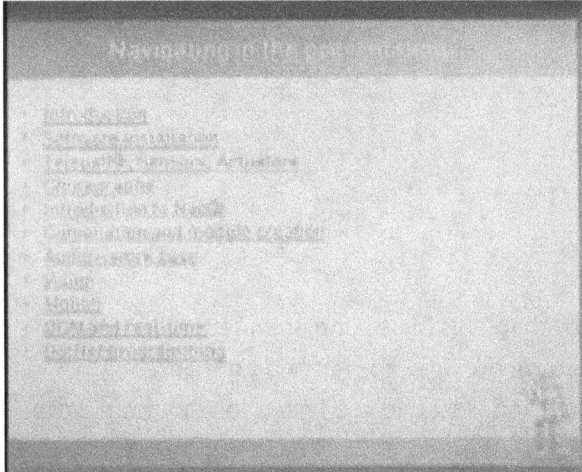

Ok the next sensor also very important sensor is the camera. Vey nice it has 2 cameras or actually i'm sorry this one is not up to date I'm not sure So we've got that Ok, so here is the new camera, sorry.

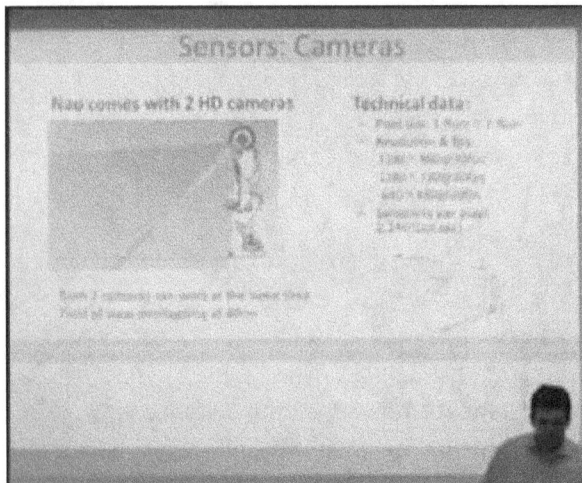

So we have 2 hd cameras, which are look at it: 1 here on a forehead and 1 in the mouth. And the 2 of field of view are crossing in 40 cm. Which means when the robot stand up you can see everything it, took out the cover, everything in front of him. From his foot, may be from here to may be 5 foot.

00:45:00

And the 2 cameras can work in the same times, so you can have a view at the same time. And the size, the resolution is {1280 to 720}, 6 megapixels. At 30 frames per second but the string is not depressed, so rule that out. Why? People make vision at the reasons. They prefer to have ruled that out, just to lock each pixel. And we have that by the means of impression. So of course this is quite huge data, very huge data, and it cannot go through the Wi-Fi network or even Ethernet. But at MTI we'll be possible to go straight around that. But again the robot doesn't have them too, yet. Probably we're walking to that. In a sense to may be integrate MTI {to enormous}. So that's why you have this, you have this resolution at this frame rate which is local data. So for people who want to make vision algorithm and write learning inside the robots head. But you can also between the resolution and have the lower resolution called, will also have a decent frame rate.

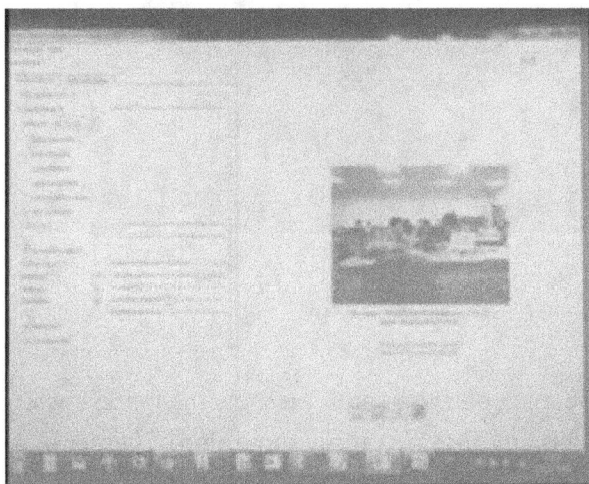

If you strip the video in your PC, I'll show you that. So this time i clean the camera better. And it's a different plugin for {Pluto}. So this time there is also to plane out, but just some settings for the camera, and put back here in red and get a bit back, get a different frame rates. You see more flashing. Yeah. Here and I can change also the resolution. So Wi-Fi is still quite slow. Probably it's decent, probably you have antenna.

So you have some settings you can change like the camera in black and white, you can also have some mark detection. Mark detection is - we provide some sort of mark, I'll show you. And you can print them, can print them and then. So it's like that, so it's kind of circular, from different sectors. So you can print them and if you show them to the NAO he will tell you this number, the number associated with mark. And this is quite fast, quite a variety. And we provide you with marks, so it's nothing to do. So it is quite fast if you want to put some marks in the environment for NAO, it is quite useful. And always you want to, if you don't want NAO to break something in specific place, you can put a specific mark on that place. -Yes! - Are we limited to the marks you give? - We provide them, different marks you can print. And, yes, you're limited to this 10 marks, and we don't provide the software. Why we'd try these 10 marks, because these are 10 best marks,

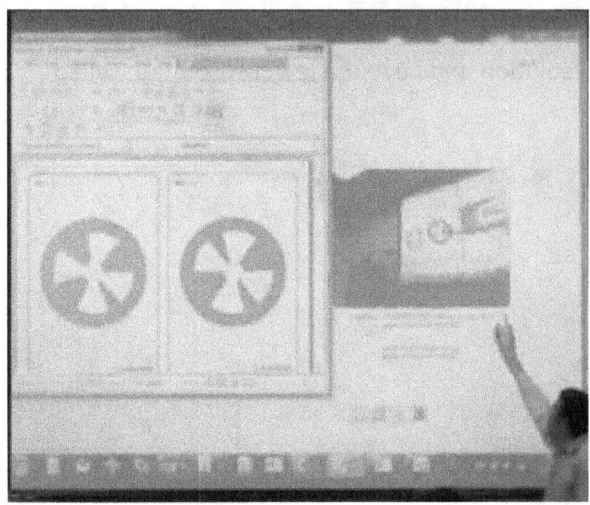

Which give the best results? So I cannot, if you check the mark detection and you show the mark, it will pull down string if he detects. I cannot really show you this, because it's usually some kind of {lessen}. But I can try, but I'm not sure, it's usually brings up last detected string.

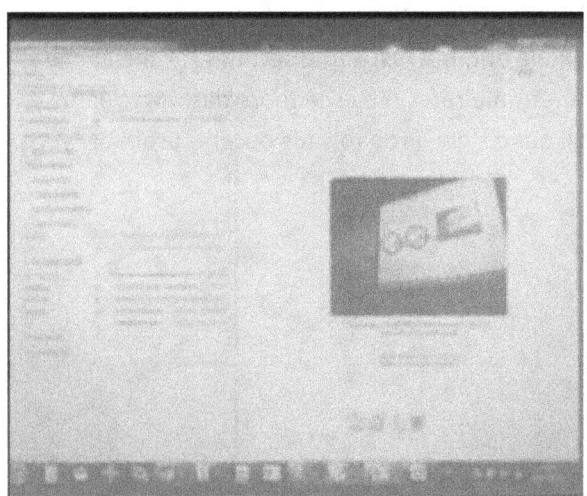

Let's try it. Yeah you send yeah the normally it can detect 0 mark and send it. And it can also work if the mark is not straight, and you margin the distance observed. Well there is a limit of course, if there are not enough pixels, where it can't detect it all. And of course the resolution is not great also here. But basically that's the mark detection. You can also use it to control NAO's detection for some cards you place in front of him. You also have the face detection, so normally this one just puts across on my nose. Look at it; it depends on the lightning conditions also. It can detect multiple faces at the same time. I'm not sure it'll detect. So it detects 3 rules, making a little bit. Of course if you're in a light room, it's better than on the sun. Am, that's the face detection. We also have a value code vision declination. So for this one you need to learn to teach NAO some object and you need to teach it to recognize the object. And then it can recognize. I'll show this this afternoon. You can that for example. But yeah i think there is no object in the database of NAO. We can of cause select bottom camera and after that some fussy cool explanation drugging this balance in change. In this camera you can also record the video. So basically this one is quite useful if you want to make a vision algorithm. If you want to develop your algorithms, you can close loop and then replay and {send it...} so you give a name, you press the record button, "Hello!" And then you stop. You can download it. So the video if recorded on now and then it is downloading to the network to your computer. So i need to connect to the robot. -All your recordings there? -Nope, it's only the video here.

I'm downloading the video on my computer. Am, so that's it basically for your plug-in on what you can check with this monitor. I'm telling you the vision recognition is quite too soon if you want in your project, you want to have NAO to recognize some object. You'll want to do that; you can go to the monitor and see it, show it. Q: {...} A: Yeas, of course this is just a tool. It's basically closed; it is just a tool without the previous 3. Yes, you probably want to do your own screen on what the MTI will provide. But for that you probably need to program in C++ to do this and also a lot of network programming I guess. And yes for you, you're supposed to imagine that you have a TV in you robots and screen, so the MTI but you turned to have a Wi-Fi, so it's just programming with Wi-Fi.

2.8. Microphones

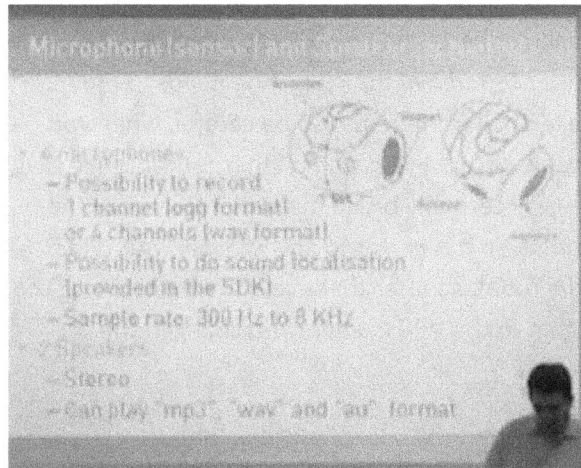

After that we have the microphone and the speakers. 2 speakers I find these speakers have quite a good sound. 4 microphones are located here, on fronts, and here, here and on backs here. So this low microphone around the head, which detects invitation and actually we'll provide this invitation. So if you just need to use it, you can just use it. And with the sound localization you can make NAO to react to any noise, for example rotate the head in the direction of the noise. You can make it walk in that direction. And you can of course record 1 channel, front channel or 4 channels, if you want. So you will be hearing every speaking and complain that it's really make you tiered of any difference in performance. That's it for the hardware.

2.9. Recommendations for use of equipment

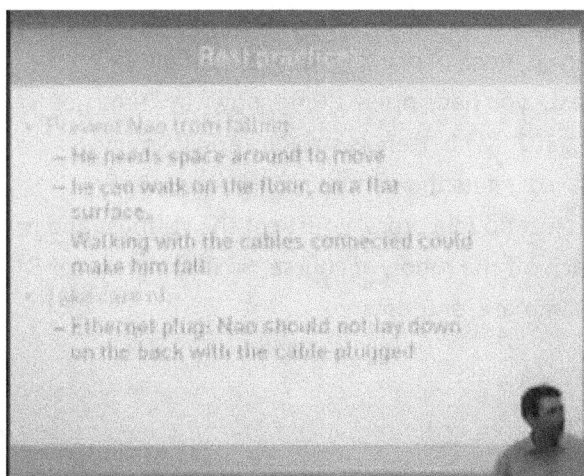

I have told you everything about hardware of NAO. Just few words about some best practices, so to evaluate now 4 need some space. Going to have great basics here and so basic spike is quite good and {...} also quite good. Usually NAO is some kind of sleepy on face, but not too sleepy. Carpet, it can walk on carpet, but depending on the whickers of the carpet it can have more or less good results. For walk, each FSR, the pressure on the foot, if the carpet is quite thick, so the pressure is not even on the sensor. NAO is trying to compensate. It has kind of strange walk. So it's better have NAO walking off the carpet - the tile, or even like the wooden floor or if the carpet is not quite thick. But also when you move it's better to be from the floor. Because be careful NAO should not fall down from the tables. Ok? The robots, if they fall down from the table they can break something. If the robot falls from its own height - usually it's ok. It'll not break itself But not from the table.

01:00:00

Also be careful pack the different cables. So you can a charger here, the thread, you may have the Ethernet cable here. So you make some demo, so it is better to remove the plug, if he needs to move. Otherwise the Ethernet plug can pull his head, finally make it fall from that. Also be careful, when you lien down the robot on the table if you have a charging cable here If you have an Ethernet cable like that and if you want to lien down the robot. If you do that you will certainly break the cable. Not really the robot, but more the cable. So you can turn the head like that, so it is better to always walk with Wi-Fi. You just need to set up the Wi-Fi once and it's saved in the robot, so after that every time you put on the robot, it can remember the configuration and connect again to the Wi-Fi, unless you changed the password of course.

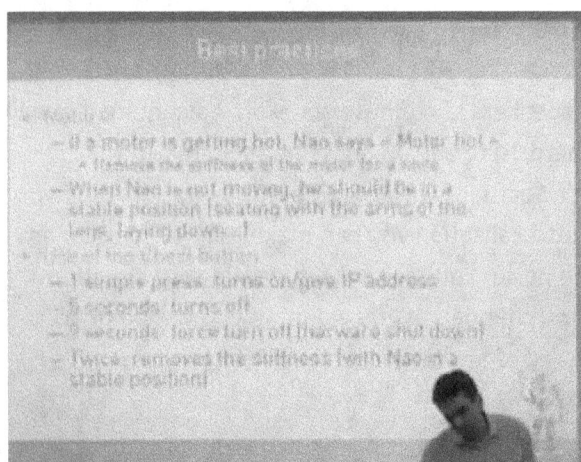

Also for the motor, if it says, NAO says something likes "my joints temperature is hot", so you can set the stiffness for her and let the motor cool down. Also lessen so much of walking, but sometimes you cannot. And also some very good practice about the chest - you have press 1 time, now say "time get rest" 1: Hello, I'm NAO. My internet address is 190.138.1.102. My battery is strong." And if you press 2 times, like that "Thank you {...}" then brings the stiffness, so it gets the stiffness. So if your robot becomes crazy or whatever a strange animation you can just double press, it just cut the stiffness and it stops movement. And you also can use the chest to switch off the robot, of course. So if you press for 5 sec to make normal shutdown. For 9 seconds to have {hibernate} shutdown.

3. Software

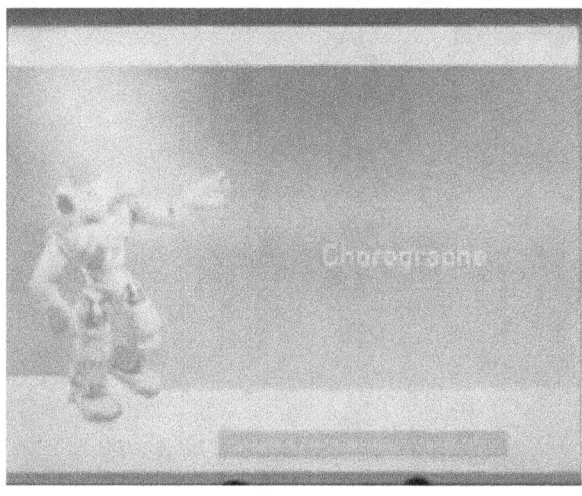

So that's it for the hardware. You have any questions? No.

3.1. Web client robot

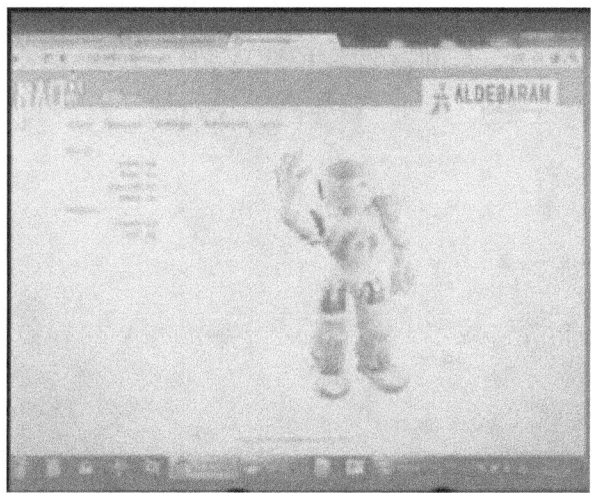

So the next thing i will talk about is NAO page. - {...} - this morning I told you about hardware and then about NAO page.

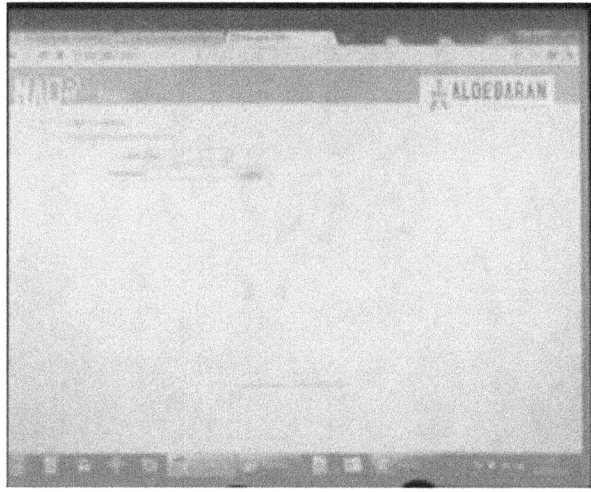

This is something we talk about for a class. Thank you very much. So like i just shown you, you just press 1 time and it give you an IP-address. You can just type an IP-address in your web-browser. You have to log in on robot - type the login and password which is nao-nao. It's the same: N-A-O

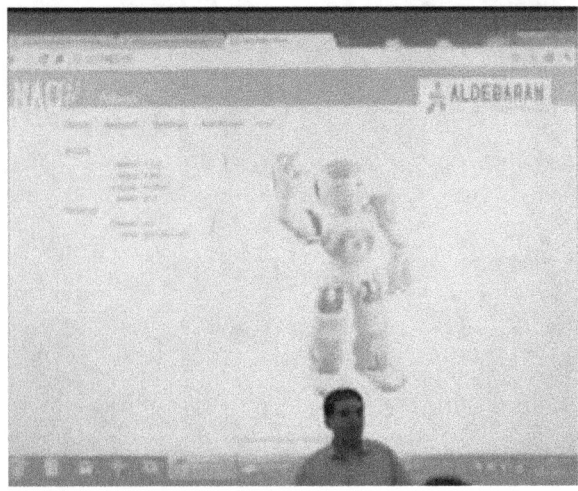

And then you enter to this web-page. So currently I'm browsing this web-page inside the, kind of, NAO. At the end {all you get...} In NAO it's Linux running, its Linux Gentoo distribution. Do you know Linux? Please hand, if you know Linux. I guess you've heard. Oh, that's a lot! Gentoo inside. So yes, you're closing this webpage. The first page, the about page can show you the current software version 4.2. And returns language and internet. I can write something, so currently only connected to Wi-Fi or Ethernet cable.

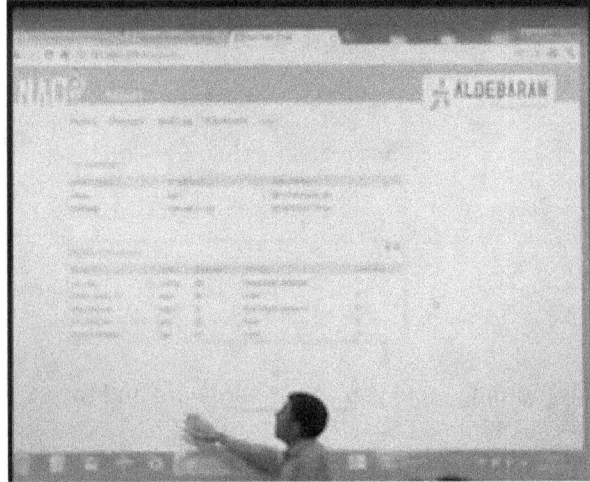

Network page. The robot lists all the Wi-Fi it can detect. So currently I have connected to this one {NSNL}, which is this Wi-Fi, yeah.

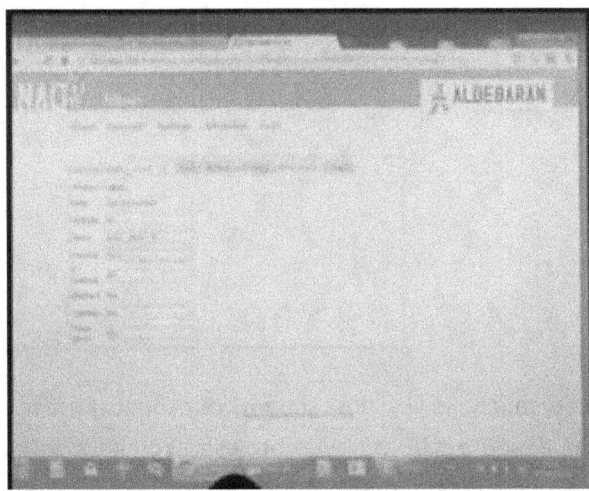

And so if you want to connect to Wi-Fi network, just need to click on it. And then you can click the connect button. Yeah, and you connect it. For example i can choose another one, this, and make it connect. And if there is a password, so you input the password. That's quite easy. You can also set this signing id if you want, which we choose this one and turn the IP-address. So that's called setting the network. And like I told you, these settings are saved in robot. So every time you put on the robot you can connect to the Wi-Fi if the rooter is on. You can also to net some settings, like hit the forget button if you don't want to connect any more to Wi-Fi network, you can hit forget.

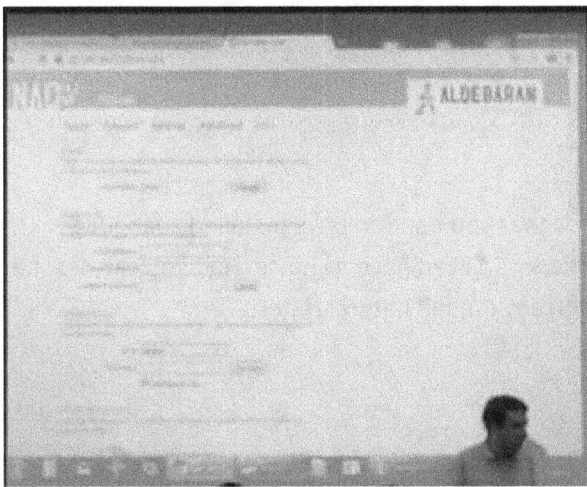

The next page there is the settings page. There is some settings you can adjust on the robot. The first thing is the name, you can change the name. That's quite useful if there is many-many robots on the same {card} that you will need.

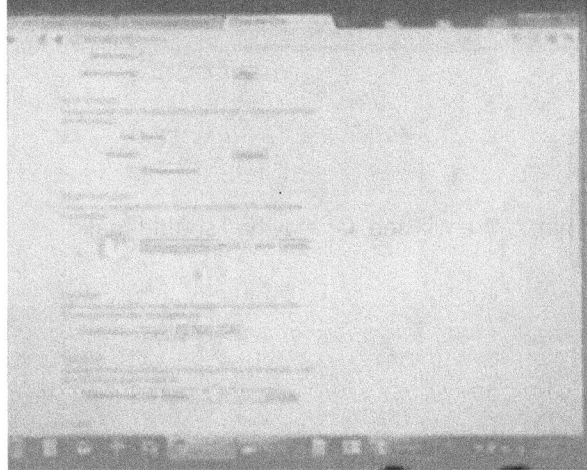

Here you would only {start to use the prelude}. You can change also the icon that is visible when you try to connect to the robot.

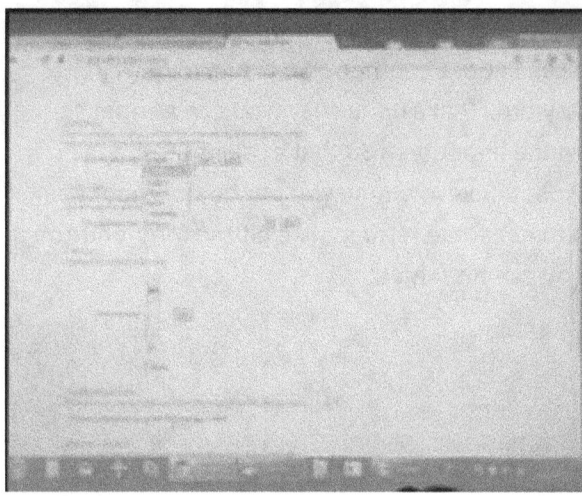

You can choose a different language on the robot, but for that you need to have installed a language package on the robot. This robot getting a language package. So i can have a choice. And you can set the time, and set the volume, and the next one is quite important the fall natural reflex.

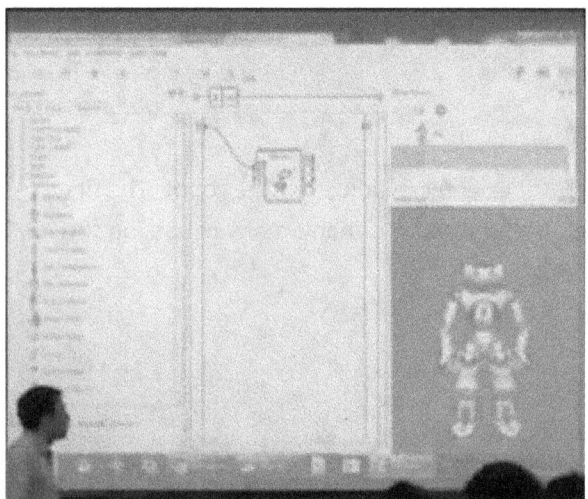

So basically we try to demonstrate it. I'm not sure it will work. We will see. Or may be I should make more before dropping it.

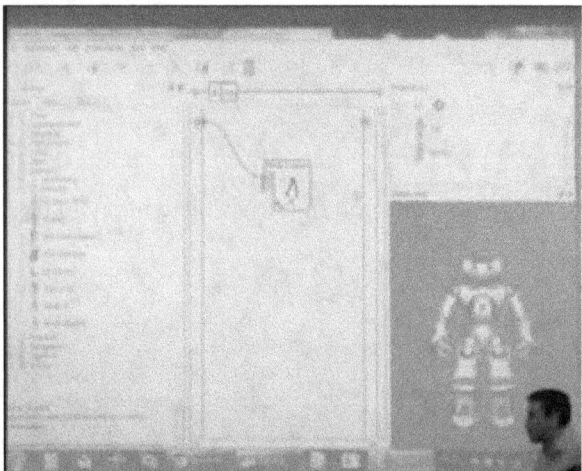

Doesn't work the network. Yes, {I can has to raster} basically, if you check this you will not try to protect itself. When it tries to protect itself by putting his stiffness and putting his hand like that.

So basically, like that or like that, depends, to avoid it to break.

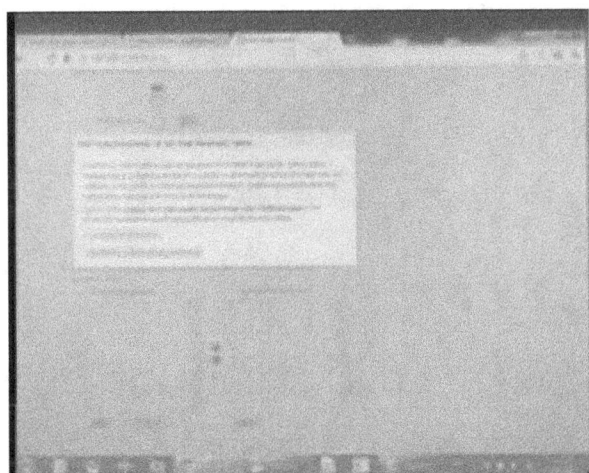

But if you check this option, then you can disable fall manager fire automatically. So if you don't want handle robot in a dangerous situation you can basically leave it like that. So that's quite cute, check if you have actually a warning, that explain you if you're taking risk with your robot, and you say "Ok, I understand risk. I really want to deactivate the fall manager".

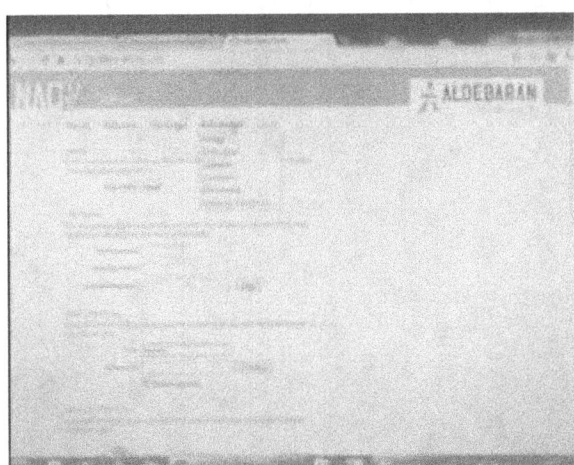

Why that? Because some user they really don't want to have the fall measure activated. Some users may want the robot to fall shortly. This for example, the example is the robot careful. For making robot careful, is the robot and how to roller, to roller jump. They make it roller jumper. But if you want the robot to protect itself {...}. The last one is the remote control.

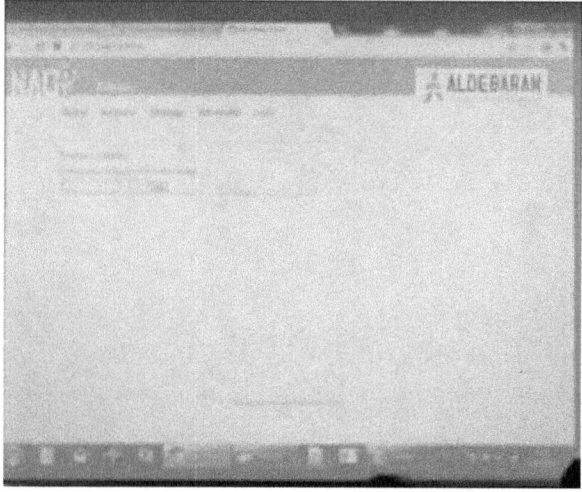

This one may be quite interesting to demonstrate. -So yeah, - Congratulations for your lecture for hardware. So you have infrared they are located in the eyes. You can control NAO with a remote control.

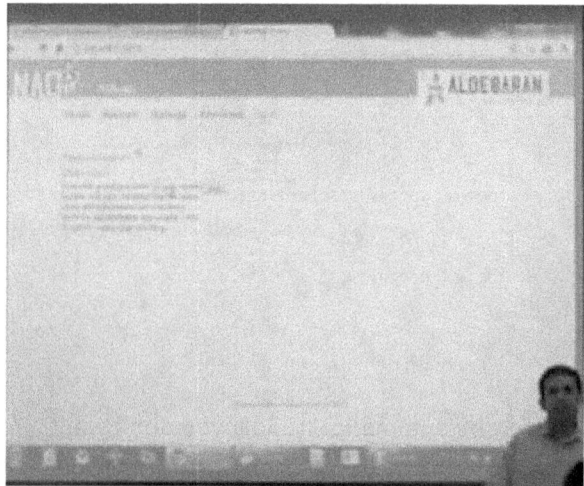

Or you can have NAO controlling device, like a TV, or temperature, or whatsoever. But to do that you need to have; you need first record your remote control. And it is in advanced page, turn directly to that, advanced robots control. We will see that statement.

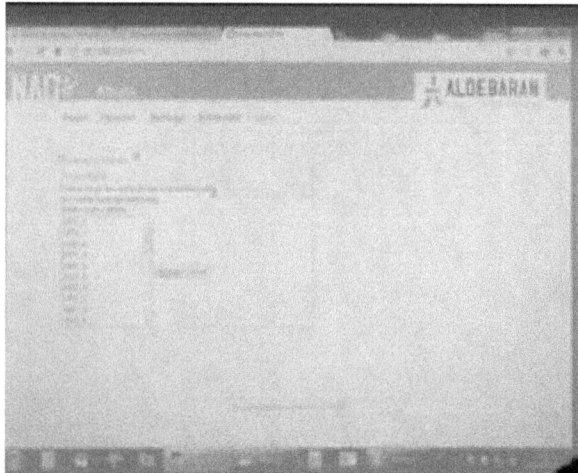

So first you need to give the name, take this one. Then get a naming itself, you can name it a new video projector. And then, so this is working only with remote control IRS protocol And IRS its open source protocol for infra-red. Ok? So I'm not sure if this controller uses IRS. Go at the start. The problem is that i need to control it, i want to control the video projector. It's usually quite cute. I have video on now. So it is a success. So because it recognized a plug control, it can record some keys. For example I would try to record "Go down". Then I'm pressing record, I'm pressing the result key on the Ok, I'm not sure goes' let me by key. Anyway, it's not very working great. It really depends on the ruler control. But you can fun keys succeeded.

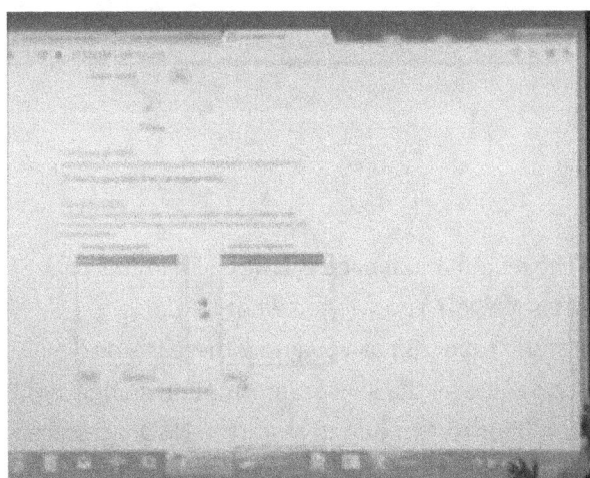

And basically after you have loaded some remote control, you see them here. So you can enable or disable all this remote control or saved pattern.

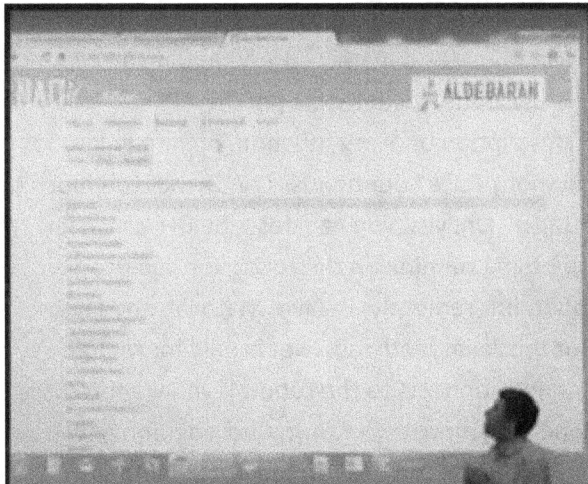

So why, basically? It's because you can record several remote control. But they all use the same {...} one euro, one key, one remotes far from another remotes. You can have confusion like that, so you can just record several patterns and just enable some of them.

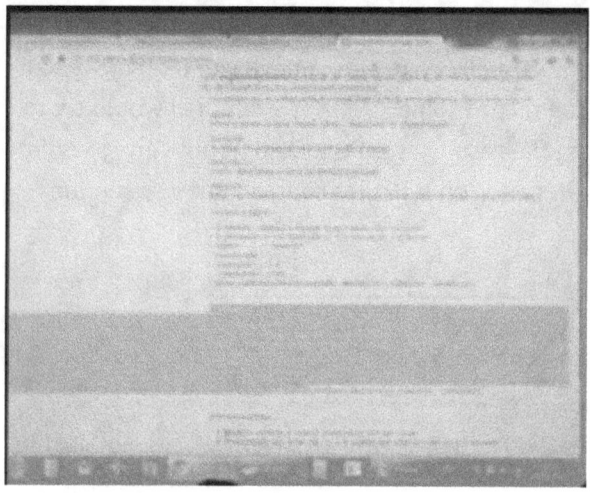

The next page, so that's it for settings, manual settings. The next manual page is NAOqi. So in this page you have a first part: you start or the restart NAOqi. What is NAOqi? You will see what NAOqi is tomorrow morning. NAOqi is the main program running in the robot. So as you know there is Linux inside. But Linux is just OS. You have main program, which handle all the sensor information going from the robot, and we call in NAOqi. So from this page you turn filter restart just to restart it. NAOqi restarts. And robot {have another look} when NAOqi restarts. I'm surprised it didn't restart. Finally restarted. So basically this page is useful if for example NAOqi crashes. So you should have some nasty behavior, stack overflow, infinite loop or whatever stop NAOqi - you can just stop it at any stage without rebooting the whole robot, without rebooting the OS. On this page you can also see all the modules inside the robot and all the behavior insides of the robot. So the modules, what we call a module is basically a {} and module is collection of methods, basically natural dll in Windows. It's dynamic library. And if you click on one of modules here, if you click on it. You can actually browse the documentation, the API documentation, of this module. So you have functional, description of every function, private fields. So for example of core Python or modules C++. You know Python? Raze your hands. 1, 2, 3 - not too much. And what about C++? Please raze your hands. C++ is better :) - Oh, yes, you can describe c++ as better. Ok, thank you Yeah, basically if you want to program something running on the robot you can use Python or you can use C++. You can also program the robot, but remotely, in Java, Math lab and java, Math lab, .Net. .Net is {java} etc. And i think that's all. But this Java, Math lab, .Net is only for remote programming. And its program running at your computer and connects to the robot. If you want to have autonomous program running on the robot, it can be done in Python or C++. Also Choreograph of cause, but Choreograph is mainly Python based, Python's program. So you can browse the documentation directly on the robot but you can also have the local documentation on your computer, where you install the SDK. The next page is the package page.

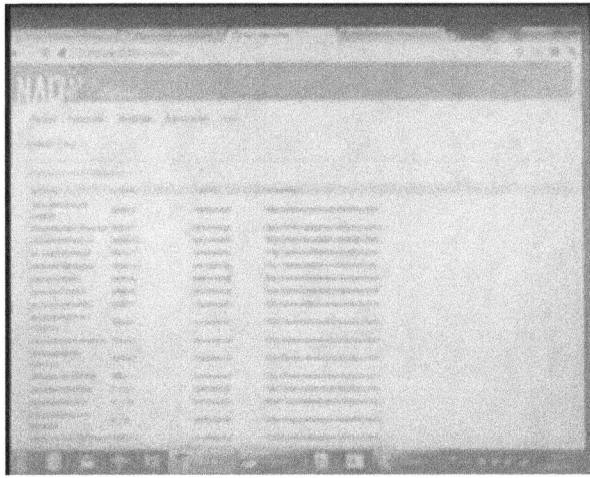

So this one lists all the packages installed in Linux.

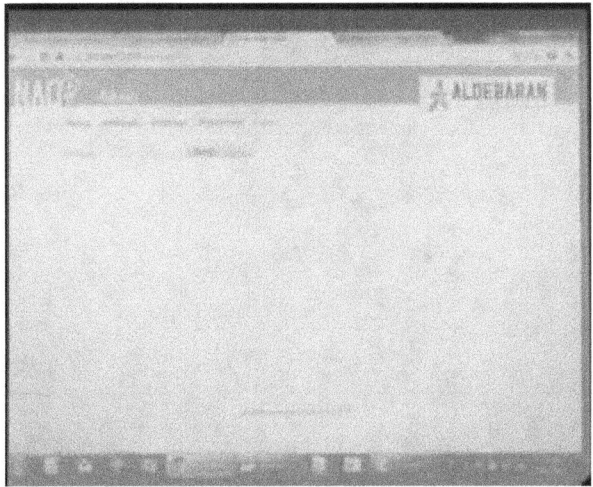

The name, the version number, the websites.

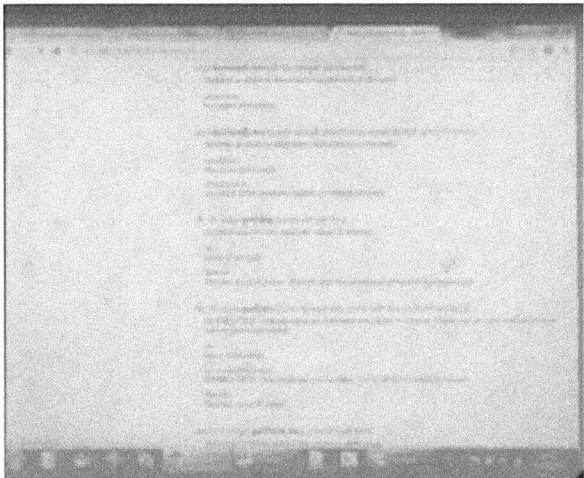

So for those who are new to these things, I'll show you so you understand what 'm talking about.

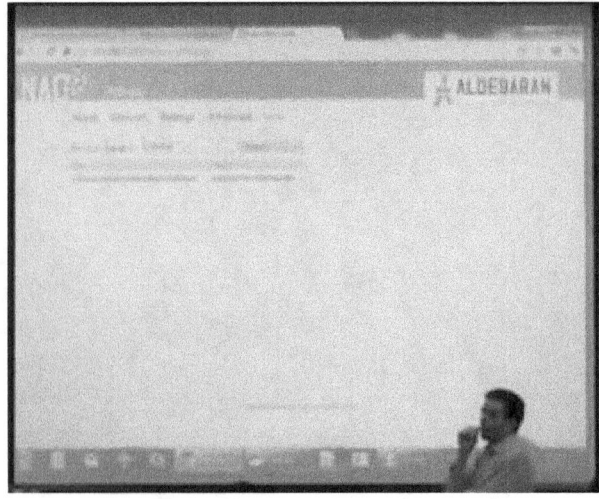

The next page is the memory page.

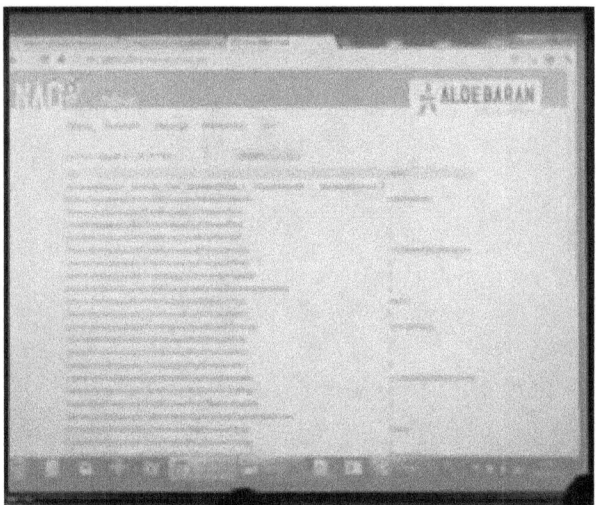

So this one is quite useful for development. So we have a specific module, which is called the ALMemory. AL means {auto learn} library. An ALMemory is one of the modules. It's here it's called ALMemory. If you look into this module you can see that there is meter like get that up, or set that up.

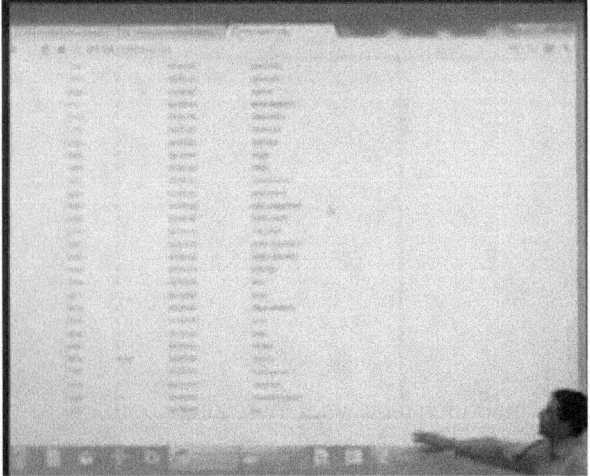

So basically memory is a dictionary, where you can install pair of key with its key values. And you can store that in ALMemory and also you can subscribe some event, some callbacks. You can subscribe callback of the ones you're interested, so if the value change then you recoded callback. So basically the memory page here let you browse into ALMemory. So you can hear a key, like for example the id, and it'll search in memory and list all the key that match this. So here so I just display my {certain number of all my callbacks}. - Is where you get data {} -So {i might not know certain features} - {..} we're calling where we are, we subscribe a whole bunch of different sensors. -Some yes, some sensors are storing data in ALMemory. Some sensor are no, but some yes. For example rest. May be I should Let me just Let me show you. It was left on the rest. So this 2 values yes, is when you press this they are changed into ALMemory. Of course, yes, that page is start it, you press when it happen. And then {.. do that when the value change} So yes, this one is quite useful. You can store on the data. You can rename. So if you make the behavior that stores the data, if you put there the value, you can look at the webpage and see the status. And the next page is the process page. So this one lists processes running on the robot on Linux. So you can see that NAOqi is running on robot, in here. You can see that Python interpreter is running also. And you have some classic running processes. The next page is hardware. So in this page you have 4 tables.

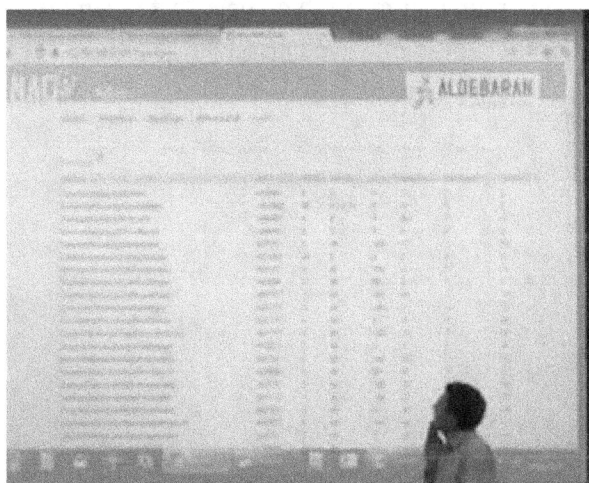

That give you some information involves the NAO. And first table list all the bones, all the motor bones of the robot. So this is not very interesting. The version number addressed by last {fantasy}.

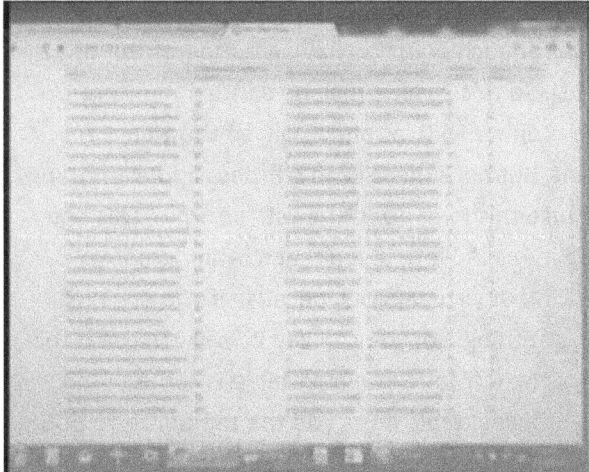

What is more interesting is this 2 column, so you have this acts and acknowledgements. So basically it tells you if the motor bone is connected, are responding.

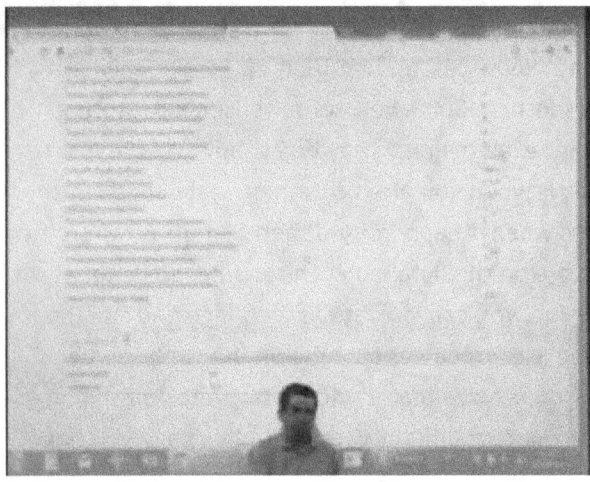

So normally it should have another act, so if it's the opposite, then that means, you have lost one more bone. In that case you can try to refresh the protocols. For that very simple you just need to press 5 sec the switch on button on the robot. But probably you would have to do this.

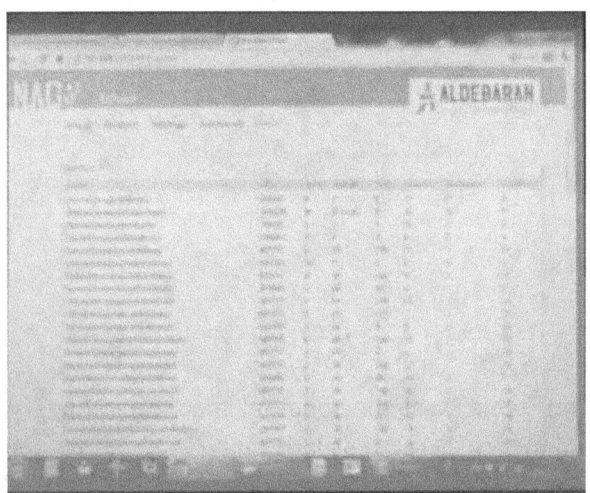

The 2nd table lists all the joints, all control. And you have for each one; you have a temperature, the 2 angles. So 1 after the gear box and the other one after {..}. So this 2 value may be a little bit different from the bottom part, bottom turning. But on top there is only 1 MRI, as i told you, so the value. You also have a stiffness ratio and electric current on the joint, but here there is no stiffness and everything is 0. You can notice this one. This one is all 0. Why? Because this joint doesn't exist. So why? This one should have only 1 joint, 1 motor. Like that, so {}. Ok? So it's called left, if you reach. And right if you reach and then 0 back. You have these joints, this some kind of virtual joints, because if you once leg, program one leg and once make it symmetrical, it'll be right thing. So, you can use it, you can use it this joint. But if you change this joint, you will actually change the left foot. The next table is some configuration in the occurrence and some {...} and finally the last table gives you the temperature of the CP. This time it'll be occurrence, different values. So basically that is all for NAO page. Do you have any questions? No, the questions? -Is there a detector, {something you can on you robot ...} -In use the camera also? So - is {...} - ah, not really. But after that it depends why you want to detect. If you want to detect someone face, then you have face detection. We also had face recognition yesterday and we told that you need the network. Actually that's not true. I did a mistake. You can make face recognition on robot. We also provide it. So face, if you have an object, if you learn object. I'll show you this afternoon how to learn object. Then you show him object it will recognize it also. For each {type of object} for that.

Easy. If you want to detect any kind of object, specter, using a sonar - all you use is camera, changing {running head something like that}

01:30:00

What kind of object recognizes {...} -Obstacle - i suggests you sonar. {..} Volume. - Does {...} -Yes, definitely. The fact that you need, yes. Ok. So if there is a data which NAO approaching - Yes? - If you need carrier of mobile phone has package have address on gather sensor and - So you want to attach a mobile phone on it? - Yes -in that case i would suggest to use and usb Bluetooth. So you can actually get just a bot. So you can buy some a cheap Bluetooth you can plug and then you can connect your mobile phone to Bluetooth. It would be the easiest way to connect mobile phone with robot. Or you can try to connect thought Wi-Fi, because your mobile phone has a gate. However we don't provide driver for that. So that means when you're Bluetooth make sure you have Linux driver for it. When invites maybe you're surprised that Bluetooth might need some driver. You just need to get sure that you also have driver for Linux, not only for windows. You see what i mean. And then you have more experimentation to do. The thing is all engineer did Bluetooth in NAO head, but we have to remove it and for stupid reason. It's because all robot is certified for the European certification. When we certified it, we asked to certify only for Wi-Fi and not for Bluetooth. So we if we put Bluetooth in it we may lose the certification for the US. Which will impacts selling the robot? So we decided to remove the Bluetooth. And cannot give you this. Even if have already put, only visitor can interact with it on their site. May be in future version we will ask again the certification people to do some. So by the way that's just connect the port, but any kind of older devices And that's it. I still have 8 minutes. The more questions?

3.2. Documentation

So before lunch i will show you NAO, i'll show you a little more the documentation you have. That's the traditional mobile phone very similar to generation of my father's :) So it had some very simple documentation in state of art. Some documentation about the software itself, so about the Monitor, where i explain to you how to use different {...} etc.

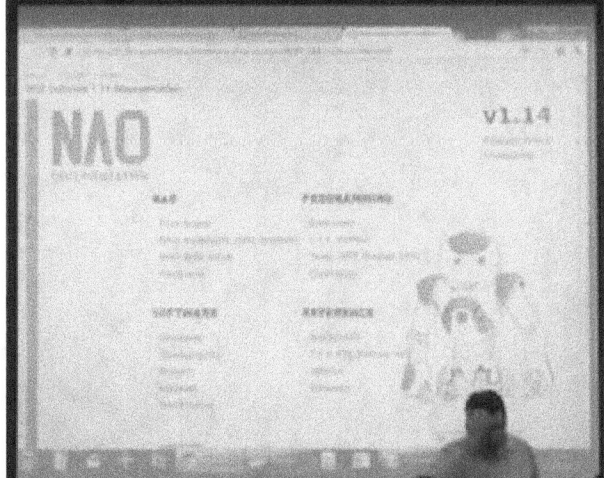

The Choreograph how to use it etc. The NAOqi. But now NAOqi you will use version 1.10 for NAO C, because we have a license issue. So maybe you wouldn't start documentation, when you wouldn't start Choreograph without you have documentation for a NAOqi with Math lab. So NAOqi runs robot with little more features, not too much more features. Now {flash of this library walk through}.

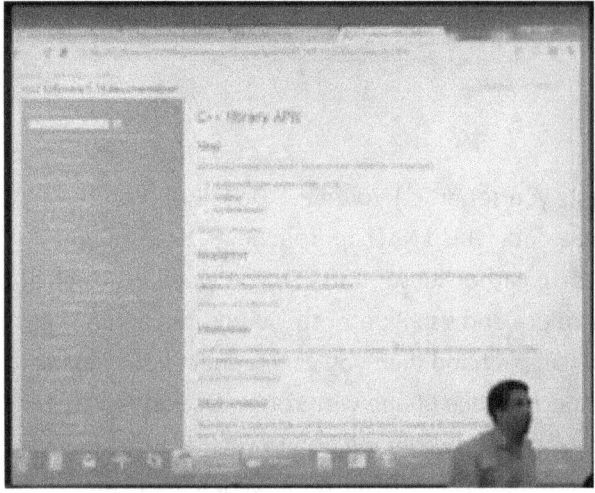

And you have more programming side, so some kind of help you programming in C++, Math Lab, in JAVA, etc. {.. distinguish}, script language {programmatic}.

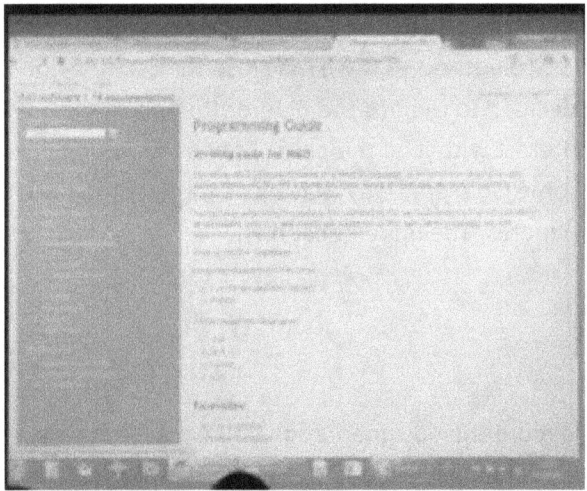

I'll show you this on the way. It's not very problematic. And you have API. So the API get {action so you and your robot}.

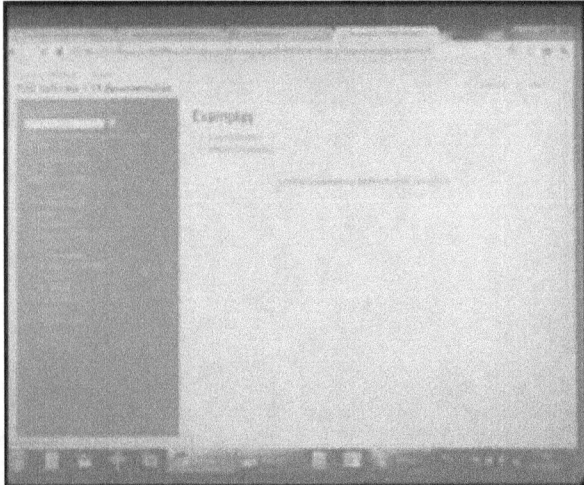

You can also of course browse it on your computer, so you can give it on the module {..} Just show you. And you can actually find a method and yes, exactly API documentation. And you also have C++ API.

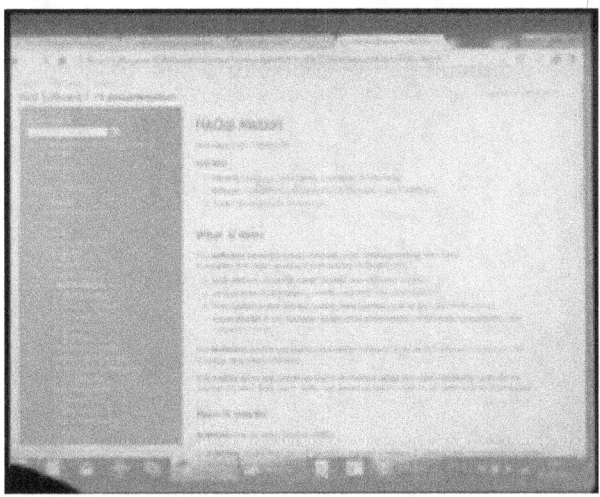

So this one is basically same API, unrelated which language you're using it has the same functional, the same parameters. Only the syntax changing with different language. But in C++ you also give some connection involved plus you programming in C++. So basically that's {probably the seed role} and the same for python, if you want to program in Python. And simply all this a tool for pressed library. Tool that we have developed for pressed library. It's a tool written in python, because engineer knew python, So basically this one is to lexicon common line tool to compile your project on different platform: Windows, Linux. You can install it, I'm not complete Python guru to be very brave, but if you have a shared folder between on your PC, Windows, Linux and want to share a folder and want to put your system here, so that you can build the same code in Windows and the same code in Linux, then ok, find something interesting to do it. I'm inviting you. Ah, as also some good, not ready to tell you about, that except of explanation of different parts of {growing robot over the ratio}.

01:40:00

Hardware use, sorry. Show you. As just some my paragraph gets lost. There is it? Oh, here. I wanted to show something like that. Because we have to see some investing method we're wishing, to have some investing method in NAO. You know what is investing method? You find a space, and NAO try to move his arm, to reach the human face. So basically starting from this point and it will compute exactly which angle or articulation how to reach this point. So that's very useful of course, if you want NAO to capture object.

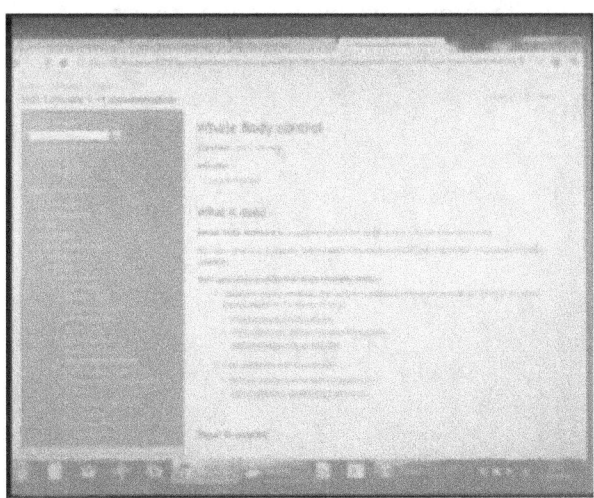

Once you know the position of the object in the world, you can just use the inverting method to say "go there" and then it'll detect it's position and go there. But the difficult part of course is to find, to locate the object, because usually you need a camera for that.

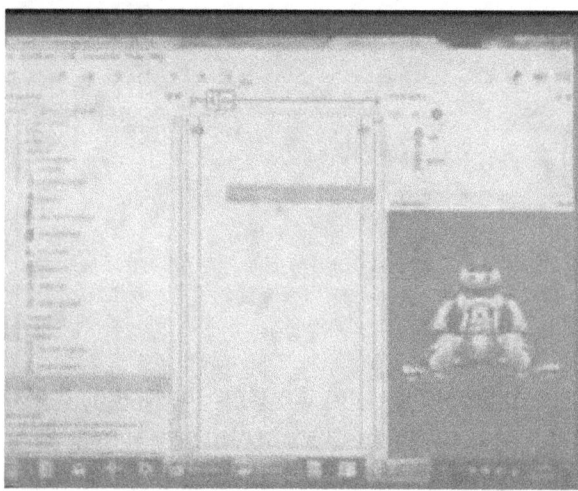

The sonar is not precise enough. Ok, there is some example of the part you noticed. And you have a kind of demo they'll show you. I just try to walk here, keep its hands. So this is just in API, you can get it in API.

And in real life it's quite useless, except for this demo. It's quite impressive, but it's quite useless. But what is more useful is investing method with a bit similar API, a bit the same.

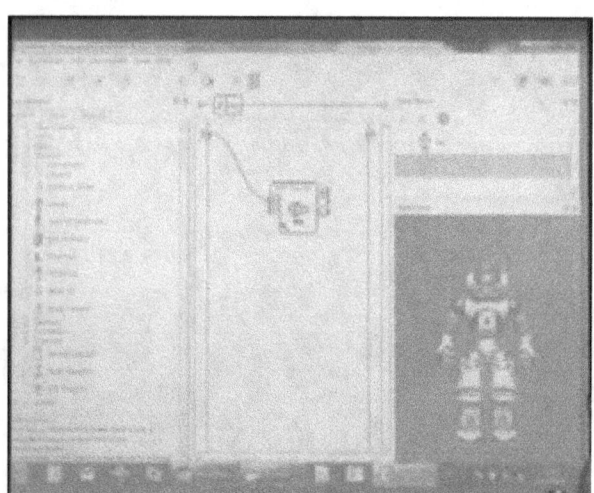

So basically here it's trying to control the head position. And so yes, that documentation is quite good, i would say. It's a lot of documentation inside. You should NAO {..} browse it and look inside and you may find what you're searching. It's not only the API. You have the API description, every function, every variable, but it also will have some example and some tutorial explanation about that. So you can really go. Any question before we go to lunch? No? No? So really we'll have lunch. And this afternoon i'll explain you Choreograph. I'll show you Choreograph.

3.3. The Program Choregraphe

So, can I start? Ok, welcome back, this afternoon I show you a Choregraphe. Choregraphe is a [...] graphical environment so that means you can make very simple [...].

Why you program that? That's because we want our robot accessible for different people, even people who don't know about programming or doesn't make software engineering. I have seen some of you are students of software engineering - which is good, because mainly we need some software programming skills to do some good applications for our robot.

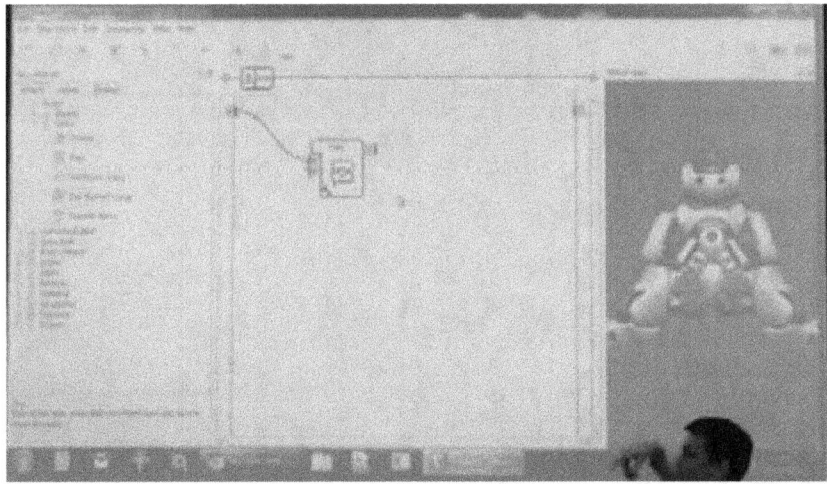

So, while you create your teams for your projects, try to get at least one programmer in your team. One programmer who knows how to programming, otherwise it could be very difficult to you. Anyway... Choregraphe is a visual programming and it's quite easy to use even for people who don't really know how to do programming.

So, first I will explain the main interface of the program, then we will go through all different boxes, that we have in box library. Basically, when you launch Choregraphe it starts with three panels like that. On the left you have Box Library, in the middle you have diagram view, this is there you create your program and on the right you have 3D-view of robot. Basically, I'm saying its visual programming environment because you can just drag and drop some little boxes like that and link the boxes together to create your program. So you just drag and drop some boxes, you make the link, and then you can make the flow. Here in the middle is diagram view; this is your program, your main program. On the left here is main entry of your program. Here, this is main entry and here this is exit of your program. The robot view is not simulator, but while you have separate simulator I will show you simulation of some sort... So it's not simulator it have no physics inside, it's just a 3D-view to give you some feedback. So, when your robot is connected you have feedback on [...] and you can also control the robot in the other direction, if you could stiffness. The stiffness button is here, I used it several times this morning, just click on it, activate the stiffness, and the button turns red. Now if you click on 3D-view you can click on several links of the robot and small window will pop-up like that, so you can control the arm like that. Just by movements. It's not very useful right now, but sometime it can have much to adjust, when you create an animation, when you need to adjust something. It's mainly useful for the hands. If you want to make an animation there you move and close the other one you can use this as a base.

00:05:00

It's a 3D-view, so that means you can pan view, you can rotate view. The next view, we have several views actually, so the next view is Pose Library.

All of views can be on dock or in different panel; different tab and you can organize all your views the way you like, except Diagram View. Diagram View is always in the center.

The next view is Pose Library. In this view you can recall different pose, by default there is three pose and you can recall all pose. It's not an animation, it's just a pose one or another like that, and this is useful if you want to quickly reach this position for the robot. Can be useful for target purpose. [1:35] time this morning, because... to use the stand. If I want to quickly put the robot in stand position, I can click on stand pose. However, if I do that... it's not an animation, if I do that it's just reach the stand pose, move all the joints to go to the stand pose, so you not stand-up. This is last pose, when you click on one pose it just trying to show to switch to reach this pose.

You can have your own pose, like, if I want to have this one, I can just click on plus button. I give the name for my pose and then I have my new pose here...

Now if I do the stiffness, the same, if I click new pose, L-pose, if I give the name [3:25].

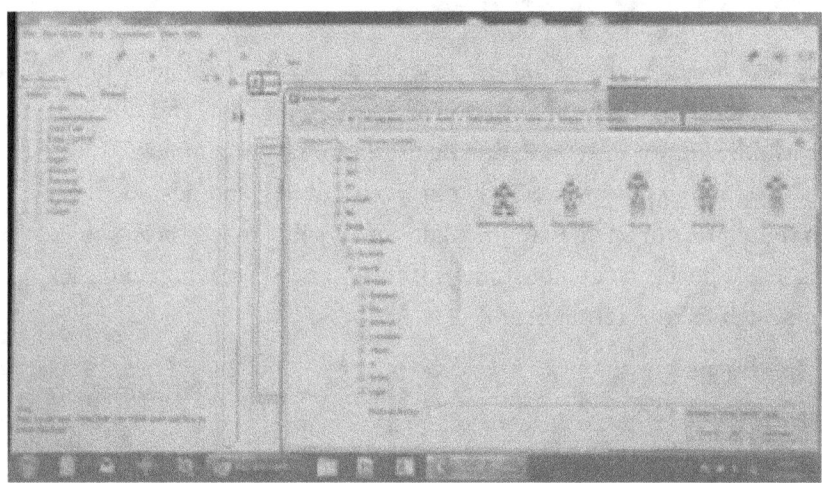

You can also reorganize your pose in different folders, just right click on it, Create Folder, and then you can move them inside...

Also add image if you really want to have some images, we provide some images, but you can create your own image. Basically, that's the Pose Library. Quick, simple library of positions. [Quiet question from listener] To create the image from the pose? I don't think the software provide that. [Another listener] No, she was asking if you can make the pose without robot in the simulator. Ah, yes, yes you can if you are just connecting... Actually, I should stop by that, because that's very important. Actually, there is... Choregraphe can connect to the robot, but also connect to the simulator. To connect Choregraphe on your PC to your robot, you usually click this button "Connect to". There Type, Name of the robot.

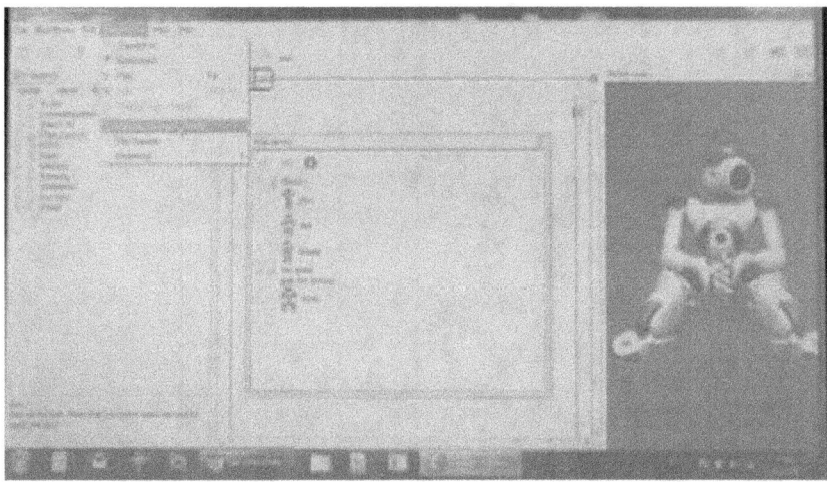

But if you want to connect to simulator then you can also choose "connect to a local NAOqi", just like that. I'm just connecting to NAOqi.

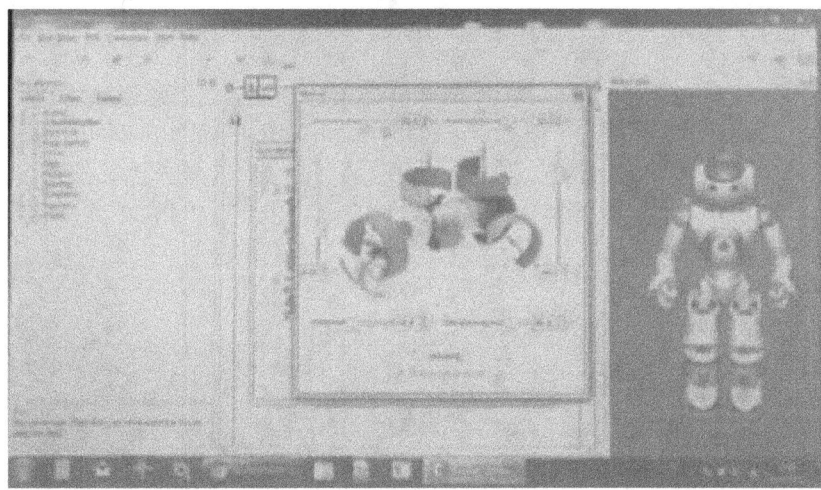

NAOqi is main programming interface. We are also providing NAOqi for Windows, NAOqi for Mac, and NAOqi for Linux. Like I just show you, you can click on one arm, for example, and move the arm. And then, of course, you can save the same way hand position.

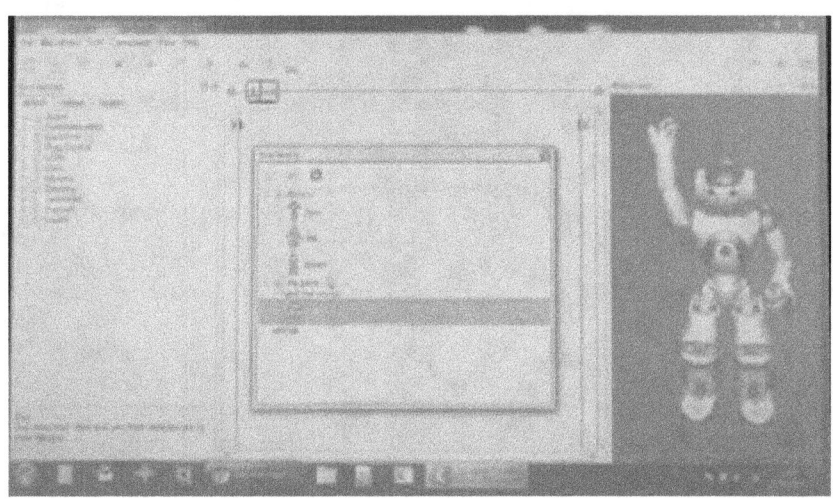

Then you have this hand exactly same animation like Hello or other. And, again, if I connect to my real robot I will able to also click on all animation... so it's exactly the same, just depend if you are connected to local one or real one, but it's all the same.

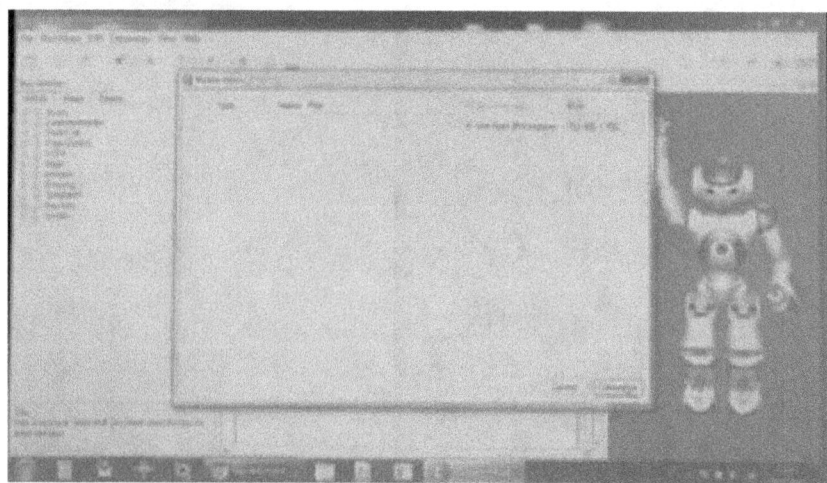

Some features, some functionality is not available in the simulator, but most of them are possible.

For example, speech recognition - you cannot do it in the simulator, even if you speak in microphone of your PC, we don't do it. Why? It's not because we could not do it, we could do it [12:10], but License problem.

For the speech to text, for speech recognition we use external software and we have license to put it inside the robot, but they don't give license to put in the computer. It would be too expensive for us to pay their prices to just give it in the computer PC.

So, for example, this feature doesn't work in the simulated robot. What else doesn't work... I think Tactile Sensor doesn't work in the simulator, but you have inertial sensory working, the FSR [?] is working, the joints - you can control them, and the [13:05]. When I said "working", I said "simulated" in the environment in the simulator. So, yes, here is Pose Library, mainly. The next view is the Video monitor. This is just feedback of the 3D robot [?], this is also working in the simulator, so we have feedback of web-camera.

You have a play/pause, and this button I use to record object. I told you this morning I will show you how to teach NAO to recognize object. So, basically, it uses this button. I can show you now if you want. I take the object, for example this robot control. I press the teaching button here, I show the object and then I can take the picture of that object, then I can imitate object with the mouse, like that. And then you can give a name, you can also give a side, because you can teach object different sides.

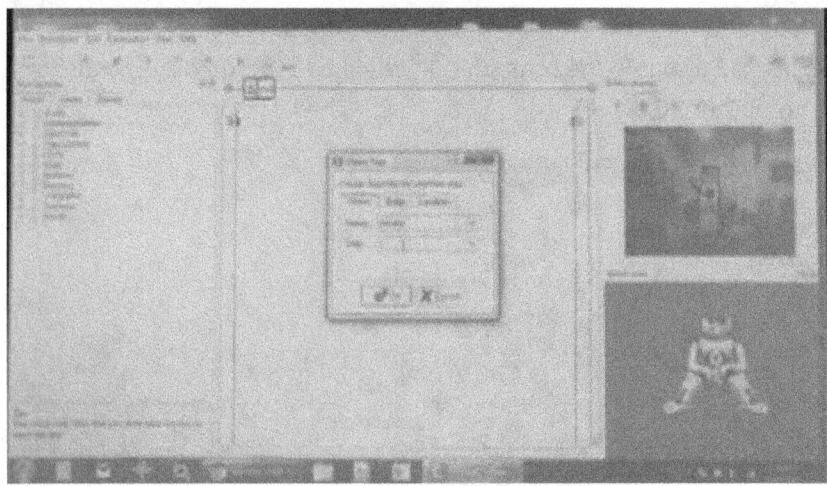

You can do that for several objects... Imitate the object...

And then, actually you just record objects locally on your computer, and then press this button to send this database to the robot. So, you send it, and the other button is just to create a new database, erase [16:20] from a scratch.

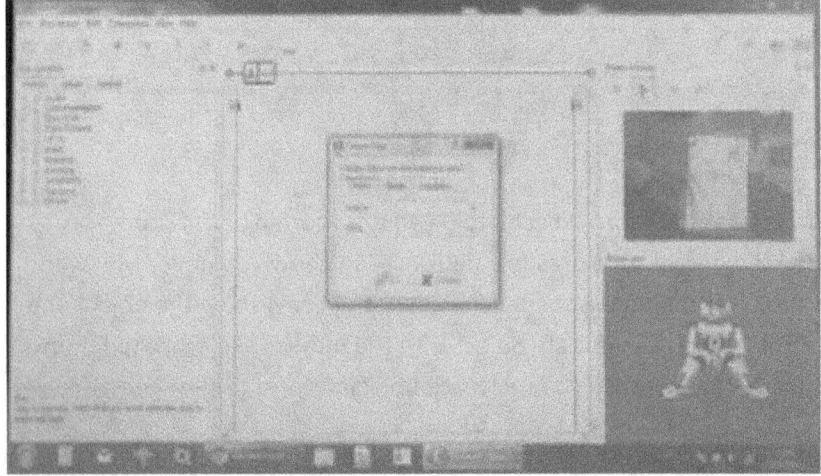

Or you can also export it and save to different folders. So later, when you have... we called the objects. [16:40] some boxes into the vision folder. One you should choose is a Vision Recognition. Just trigger the first input in the box, and here the output which gives you the object which now is recognized.

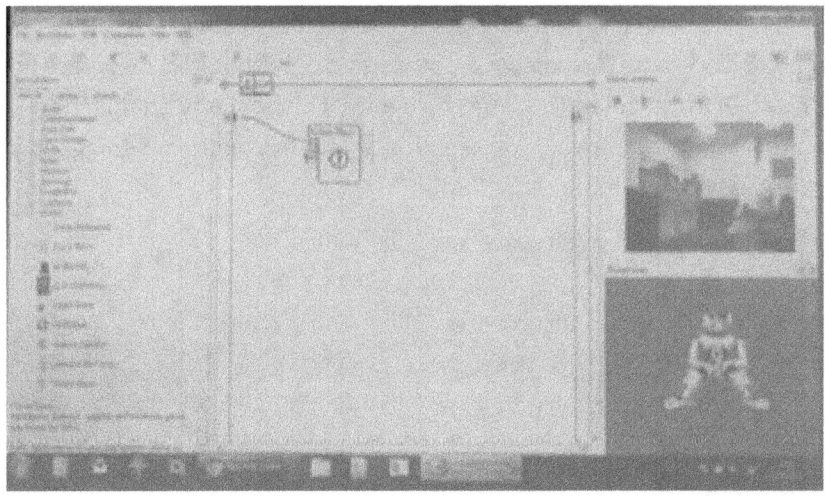

And we asking to say it. I don't take this "Say" button, because this "Say" button is just says "hello".

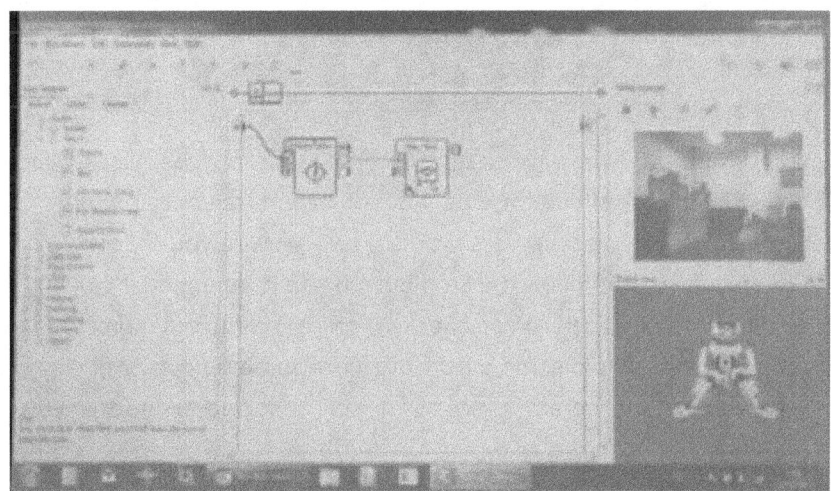

I take this "Say Text" box which I copy and paste at a back level.

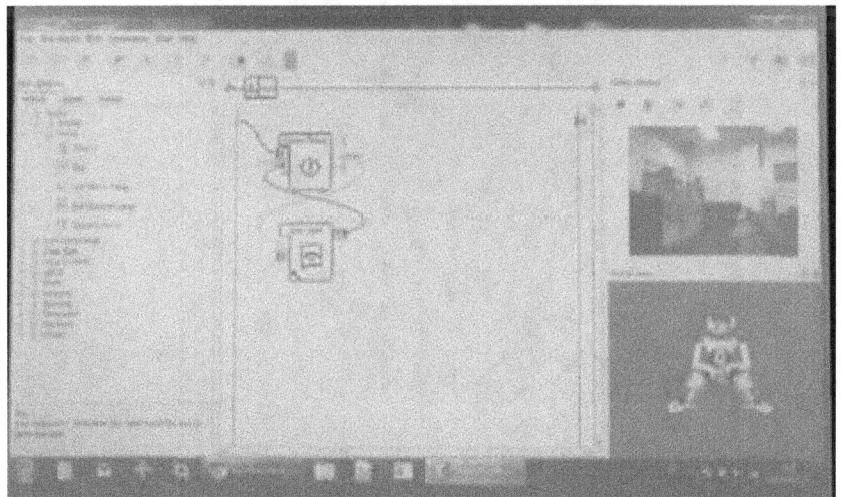

Will expand that to be tethered. I just want to make a quick demo of object recognition. I linked the object recognize to the input of "say text", so we say object name and side. Let's try it. [Robot voice] "Remote front, remote front..." "Book front, book front..." It recognizes it several times and just says it several times. If you want to avoid that, once you have recognized object you can stop the box. So it is stop the box itself. And by saying object you can restart box. [18:35] [robot voice] "book front, remote front" It works no matter which direction you give to it... and the distance also. I didn't like recognition

[?] and the object itself. And also, the way it's working - is it's using the texture of the object. It's just paint the picture [19:17] imitate, to try to find some interesting pic samples. So, I cannot [?] recognize or white or black paper. If your object is too uniform color, it will not recognize it. That's why on this there is menu button, to recognize it. If it's too small or far away, you see - it's will not recognize anything. It is also good, when you learn object, to try to use maximum... the frame.

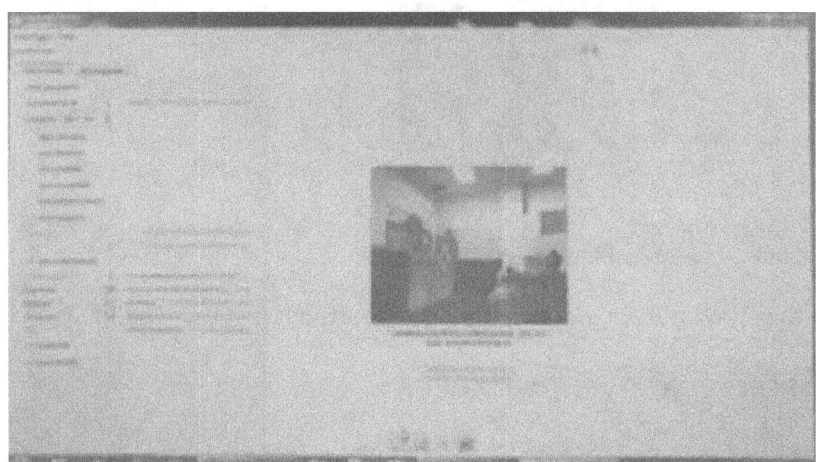

And, by the way, if you want to use that, I think, you will have to test it on the real robot. Because in the simulator you cannot bring the real object inside the simulator. But, yes, you can make your program and just assume that you will have it later from time to time. This morning I... I showed you the vision recognition and use it. Program is still running, so I can press the Stop button and it's stopped. In a Monitor, normally, this is, you can see the remote [21:18] so you check vision recognition in monitor. Basically, this is it for the Video Monitor, play/pause and some buttons to learn some objects. Video monitor is also useful then you want to debug your program, especially if you use face detection, so you want to see the bound it's see.

The next view is Project Content. When you save Choregraphe file, by default it's save CRG-file. Actually, CRG-file is just a ZIP file, that contains different folders and also some xml files and contains some python scripts inside, but they zipped in one file so that you can easily copy the Choregraphe file to other computers.

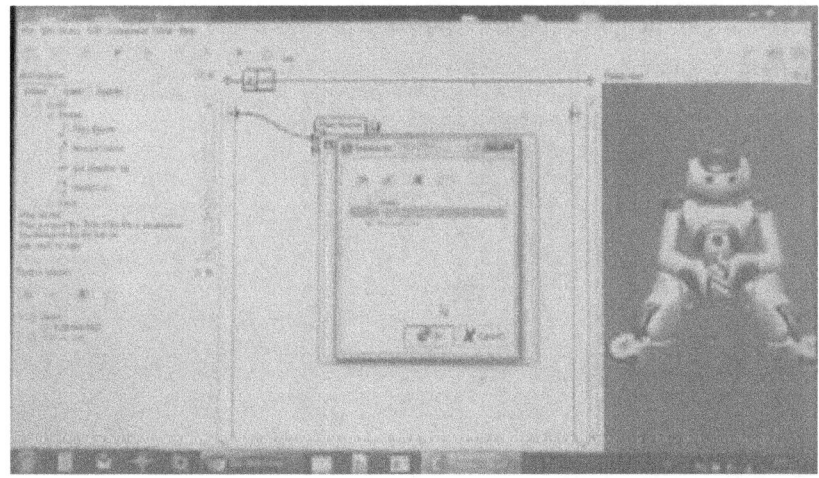

Here, with this Project Content you can add some files inside your Choregraphe project file.

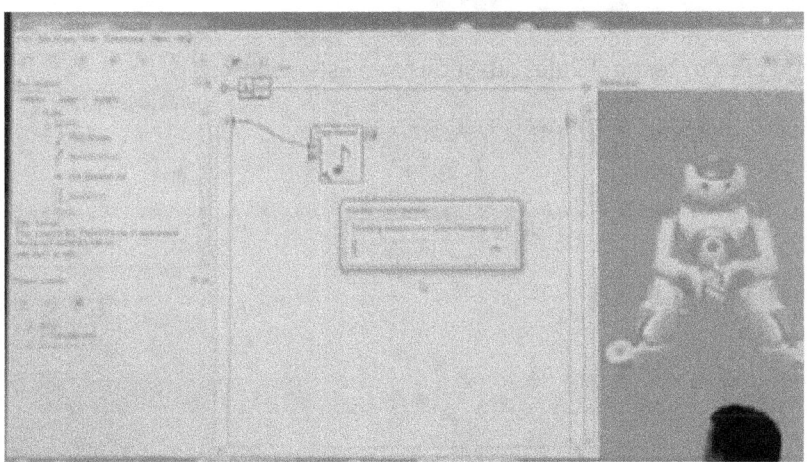

For example, you can add some music, so if you want to have some mp3 files, and you can also create some folder to organize it better. Your files... moved some files into music folder. And then I will test it by using "plays sound" box. We can click on the wrench to set parameters, and one of the parameters is the file you want to play. You can type it directly, you can also click on Browse button and it actually display same window as Project content where you can pick up file, this one, OK, and you can test the program.

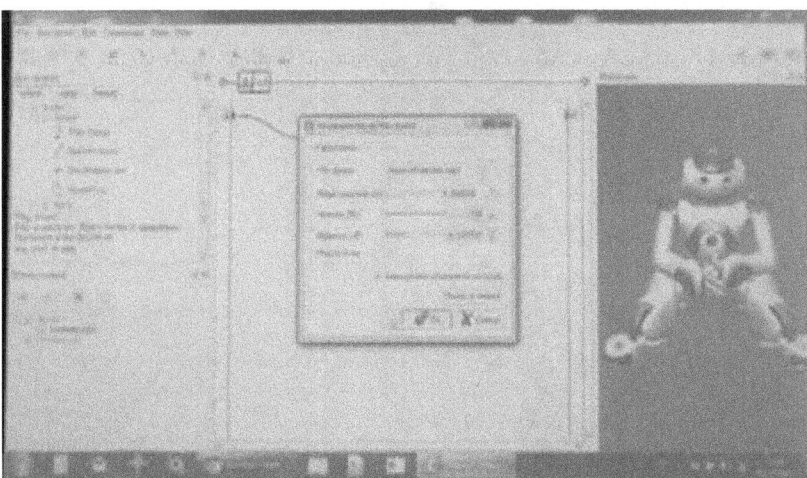

It will download all the Choregraphe files, also downloading mp3 files. [quiet question from listener] Yes, when you press play it actually downloading file from temporary folder, but you can train many application like that, many program. [24:55] in the good place, in the good position, not in the temporary folder. This play button is mainly to

So you just use it in your program and click to test it. [Quiet question from listener] Exactly.

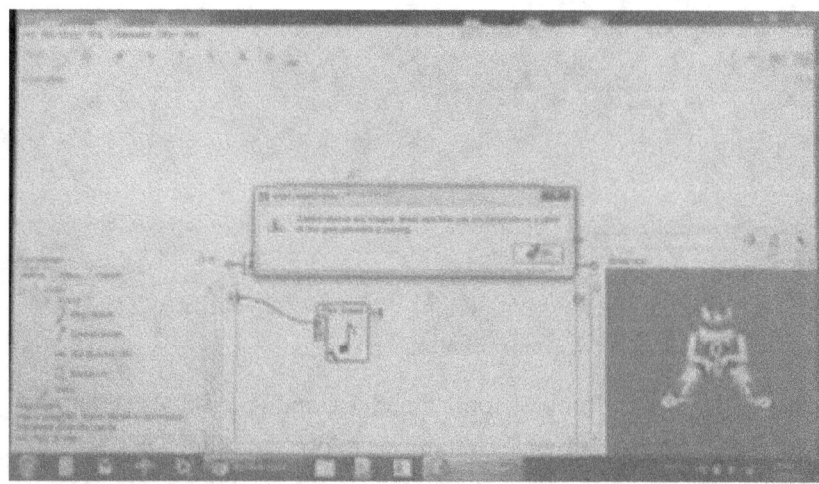

Basically this is just to use Project Content. Be careful not to leave to many useless files, because it takes more time to test - its use all the files every time.

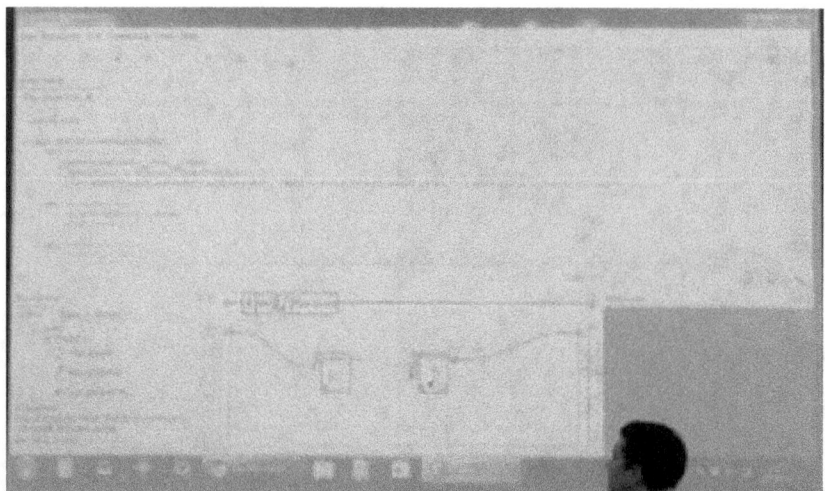

Just use some necessary files here. But for you, for simulated robot it's not a big deal, because when you will press play it actually copy file to the folder on your computer, because your simulated robot running on your computer. So it's just copy from one place on your computer to another. It's usually quite fast; it won't go through the network, so it's no big deal. The next view is the Script editor. In this view you can edit some scripts, python scripts for some boxes. Because there is different types of boxes and some boxes can be script box.

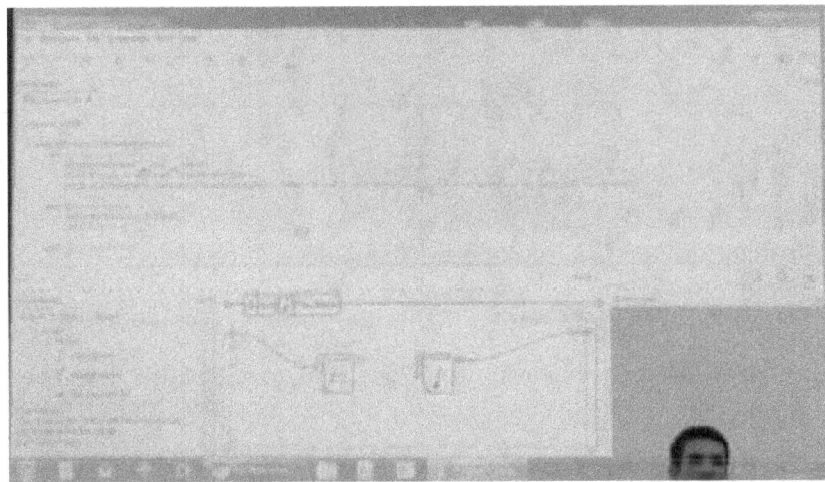

I'll show you example of that. For example if you click this one, then you can see some python script. You can type some python script here in this editor. Choregraphe is quite limited editor, so you prefer to write code in a more elaborated editor, after you can copy-paste inside this box. Or you can just make some simple box, that load and run some python script which you add into your project content, it's also possible.

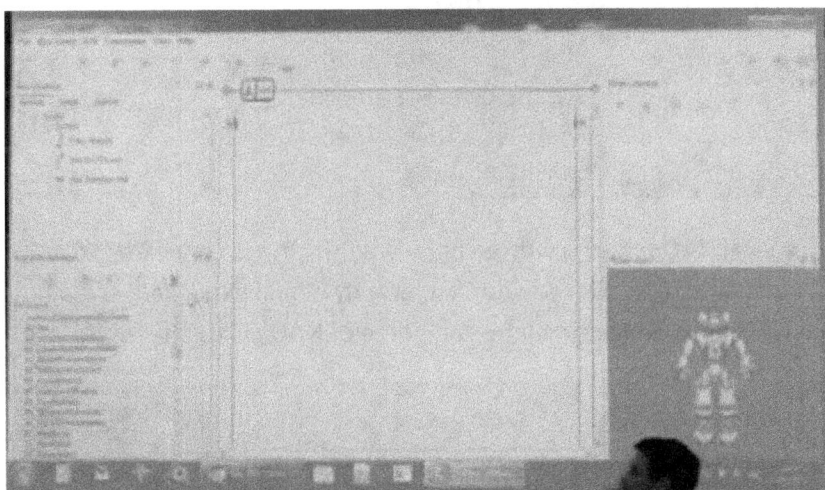

In the project content you can add any kind of files. It can be script files, even compiled library if you have C++ module, your own C++ module, you copy it, compile it and put it into your project content, and load it from program. The next view is Behavior manager. And this will actually answer your question. Behavior manager is a view you can use to install some behavior, some program in the robot. And of course you can remove it. Basically, you can click this button to upload behavior to the robot, like this one. No, this one is to upload current behavior which is empty.

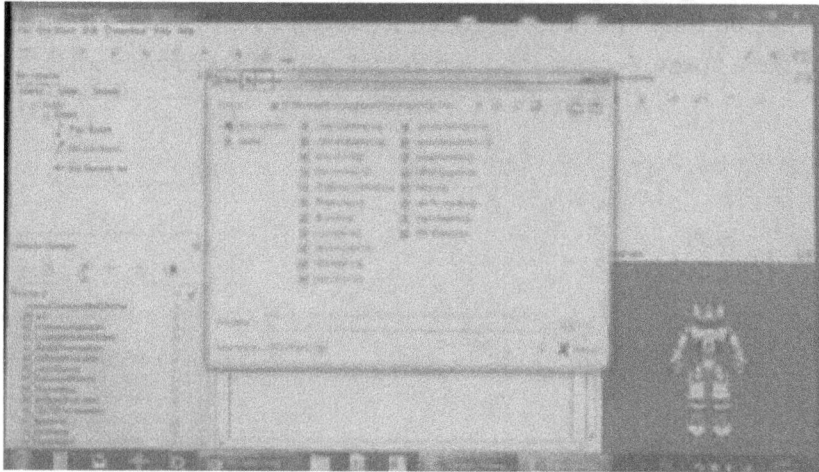

It asked me a name for it. But this one, you can choose behavior from your hard drive and then it will be downloaded.

Here, as you can see, I connected to a local NAOqi that is written here in a title, if you connected to a local one or if you are connected to a real one. And then you do that, exactly what I explained, actually brought behavior to a folder on your hard drive which simulates the home of NAO inside the robot.

That's why I have many applications in this folder. Here I also have many applications like the one I showed you. In this folder you can also press the play button, there is little play button here.

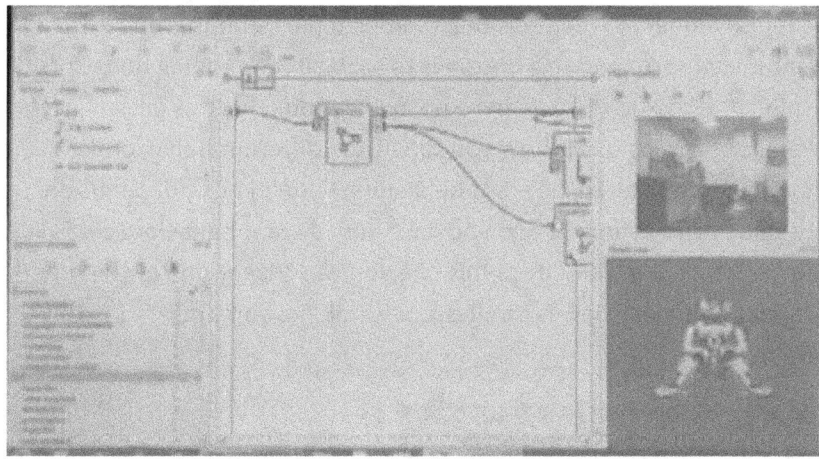

If you just press it, it will run the program. And here, this checkmark, if you check it, then that means this program will be launched then the robot boots.

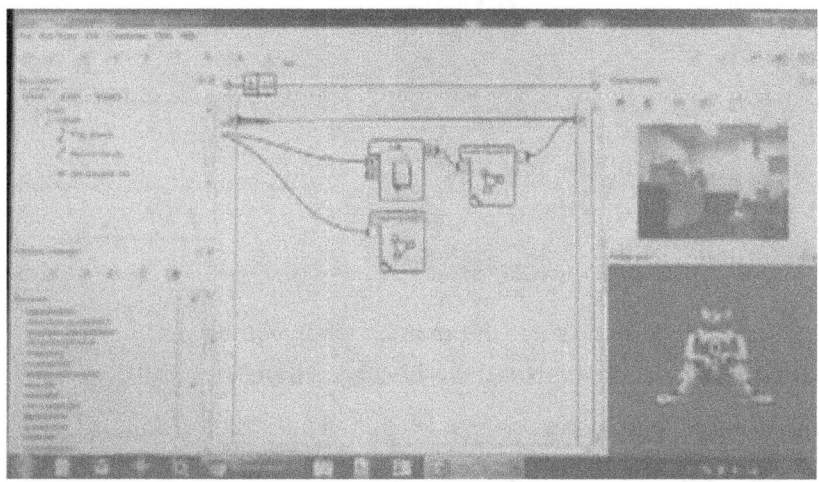

Every time NAOqi is started it will launch all the programs checked like this. Here on my robot I haven't launch any programs that's why I'm continuing to logging [?] but if I want to interact with you, I usually launch the main program which is NAOdrive, this one.

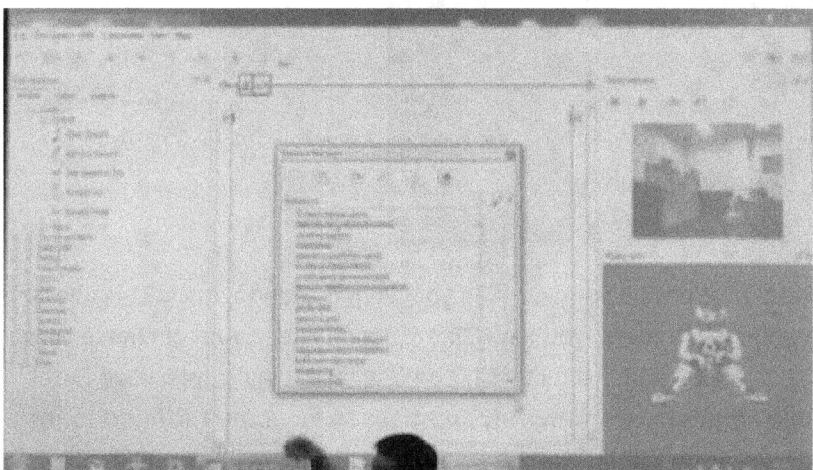

I'm just checking it, and then every time I restart, every time I read - NAOdrive is started. [Question from listener] Can we see NAOdrive? Yes, you can download it even after uploading it to the robot; you can also get it back. If I select it, I can just choose "Open" it will download the file in the temporary folder on your computer and open it, or I can save it on your computer, so you save it somewhere. So, yes, if you want I can open it. This is for download, saving, this one to upload files, and this one is to delete file. Delete file if you don't want it anymore. And this button is to stop all the running behaviors in the robot. Currently, there is no running behavior. [...] You have no application... for this project, I mean. That's mainly the Behavior manager for you; this is how you can install different behaviors. Usually it's a good way to try to split your application to different behaviors. I don't know, depends on you, but if you have great complexity value in your project and you have very distant behaviors, and have just one main behavior that you launch at start, and from this one you can trigger one or the other - it's better to organize your code like that. The next view is the Resource... oh no, sorry doesn't exist in the one ours. This is 3D-view, we have seen it.

The next view is the Debug Window. This is just to show some information when you have some error messages. Then you run your program if you have some errors it will be logged here.

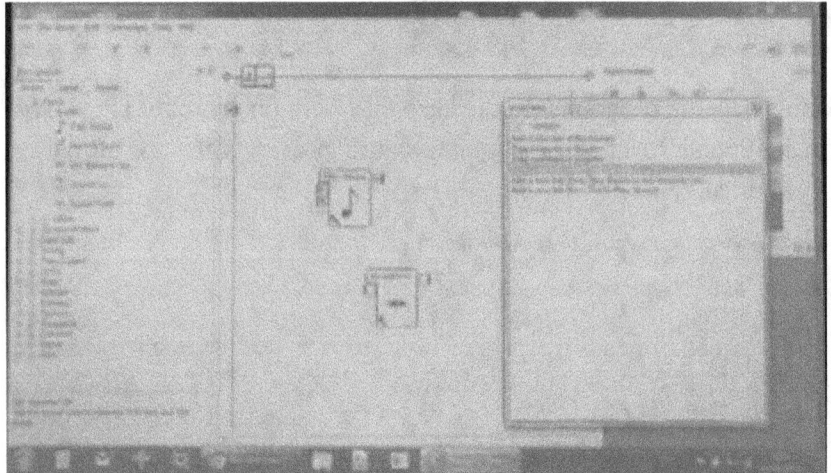

Or you can even log it also somewhere, information yourself. And you have the Undo Stack. This one is just to, if you want to undo. For example, if I do that... you have all the actions here. And of course I can type Ctrl-Z to undelete or Ctrl-Y to delete or I can also directly click on one step. Finally, you have Reset Views. It just reset default three panels. That's it for a view. Any questions so far about the interface of program? [Question from listener] The script editor keeps recording on insert, right? Yes, you don't intercept... If you press "play" it will save the files. After that, what else we have - the simple help menu, where you can access the API.

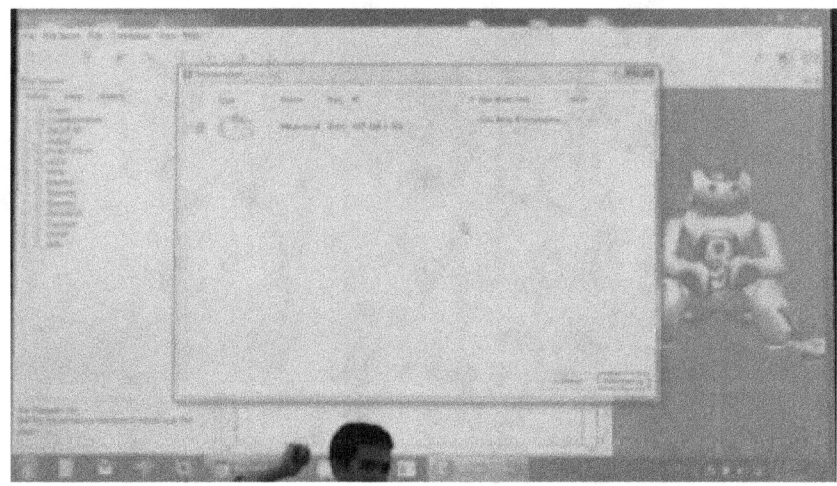

Again, that's exactly the same kind of API, very similar to what I showed you this morning. In a Connection Menu, I already showed you how to connect... sometimes your robot aren't listed here. Actually to have the robot here, we use the Bonjour protocol for Mac add-on. If Bonjour protocol is installed on your computer, normally it installed, and then you see the robot on the same network. And you can just double click on it to open. Select it and choose "connect". But sometimes this bonjour connection is too slow to establish. So you can also directly type here this is everything you have to know about connection. You can, of course, disconnect from a robot. To disconnect I have nothing. Trying to reconnect to a robot. After that you have some other: Play, Stop and debug output - this is the same as this button, so some shortcut - connect, play, and stop. These just show debug output, it's exactly the same as this window. It's just a quick shortcut to see the debug information. And this one is to connect to local NAOqi, simulated one.

Advanced, that is why you will not need to use it, that's to flash the robot with the new version of the OS. So do not. And the File Transfer. The file transfer, you can actually load into the home of the NAO. Again, the login and password is nao, nao, just like for the NAO page.

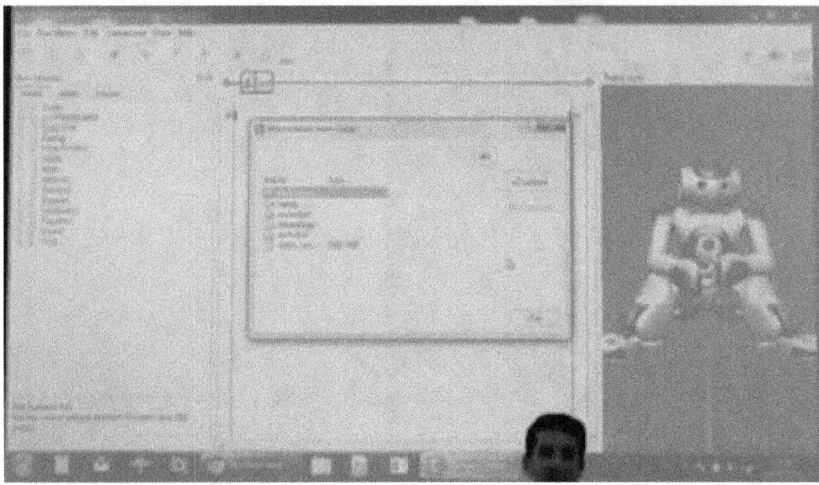

Then it's just display the folders which are inside of the home of the Linux OS. It's really home Nao here.

3.4. Development behavior blocks

So you can easily upload some files from your computer to the robot. This feature is not working if it's the simulated robot. It's only useful to transfer files to the real robot.

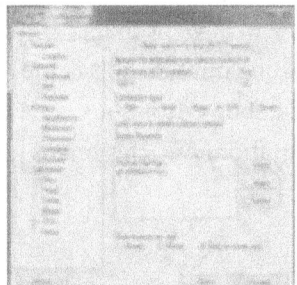

[Quiet question from listener] Not in this one... but you can do it in SSH, actually.

Nao user. Nao, nao - password [question from listener] ...

completion of the request, some special binary you need install? No, don't think so. Depends, if you want to use self-auto hybrid... to install binaries... compile it... mainly you need SSH... [From listener] So, it is possible? Yes, it is possible. [Quiet question from listener] Yes.

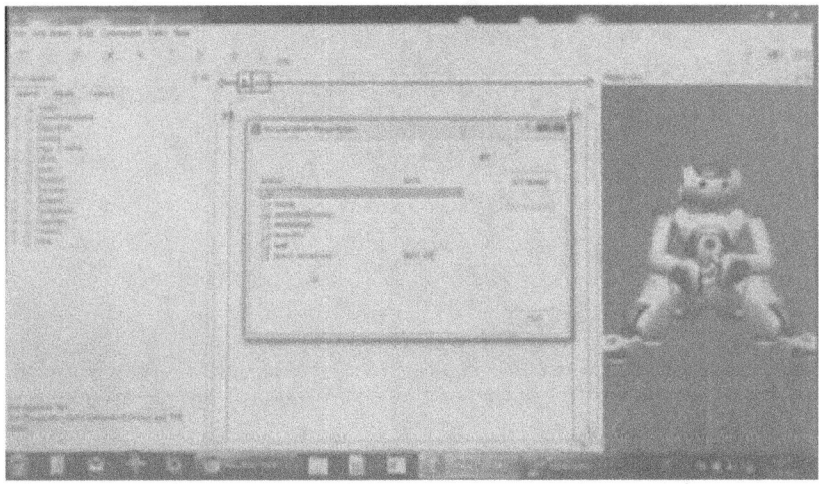

Here I'm in the home nao, I can just show to use super user, and password is root, just root. And then I just want to exit the root level. I just want to show you, even if you are just the normal user, you can access the root level... So, here I'm at the root level, and if you want to go back to your home... So, you have the same folder as what you just saw in to Choregraphe file transfer.

Exactly the same files. You can delete files; you can create folders, the same. Created test folder here, and then test folder look here, really the same. But in Choregraphe it's limited to the home, so you cannot do strange stuff in the OS. If you ssh to it, you can do elsewhere, sometimes you need administrative rights, of course, then can do SU and the root password, and then you can install whatever you want, driver or... [Quiet question from listener] Sudo, yes, it works, yes... Yes, I'm thinking about, mainly for you... talk about, the root was long gone (? No go?), so if you want to put root... now you talk about some form, I suggest Bluetooth. If you have good driver, you probably need to install driver in the robot connect. But, of course you cannot do that in the simulated robot, for you it becomes more complex. I would suggest for your project try to find something quite easy, because you mainly work with simulator, you only have one real robot for all the teams. If you think of device and link it... it would be difficult for you during the project.

Basically, it's very simple interface to upload quickly some files, if you need to, and that's it for Connection. In Edit you just have Undo, Redo and some Preferences for the application. Just like the Language, nothing very important, actually.

Maybe this one is useful for you. If you decide to work with different NAOqi, so if you want to have a test version, another version of your simulated robot, then here you can specify where the folder of NAOqi on your computer is. This one is for the simulated NAOqi, so you can specify which one you get, and this specify which version of the robot should display in 3D-view then you are connected to local NAOqi.

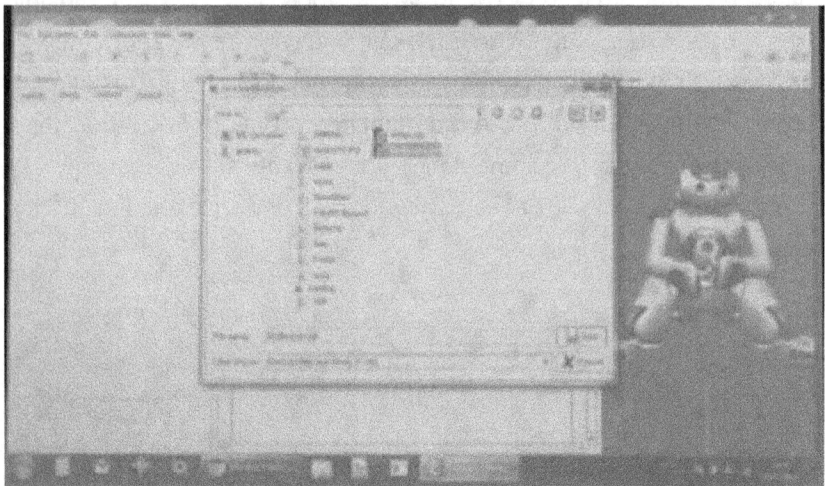

But you can leave it blank and it would find default folder. That's for Edit. Before Box Library I go to file, nothing special, just New Project, Open Project, Save and Exit and Box Library, finally. It's about here. Because the default box library we provide is read-only.

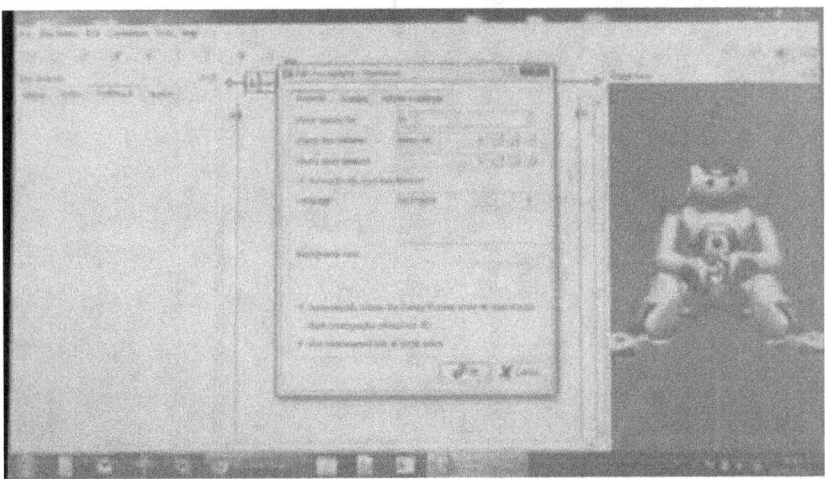

You cannot add, modify box. But you can create your own box library. To do that, you just go to the Box item menu and you choose new box library, and it created new tab here, which is currently untitled. And if you try to save... I should rename it like that, I can't do that. Rename, no, not this one. I should Save file, yes, you can type it... this one, place it. And then, Save As, and then you could rename your box library.

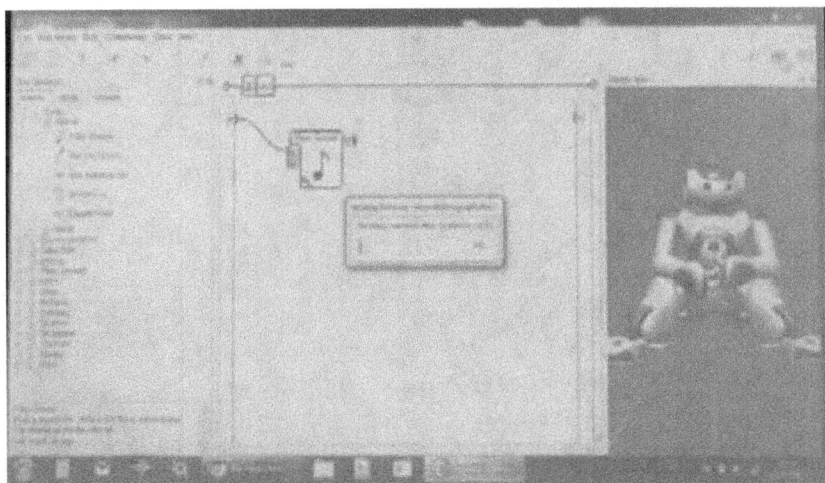

Then you can add your own box. If you create your own box, you can add it and group all your boxes in your own library. And if you want to load this box library every time you launch Choregraphe, you can do it in the preferences. In the preferences here, user's box libraries, you can specify your box library that you want to open every time you launch Choregraphe. You can specify so box library. We close it, we don't need it anymore, I just have mine, and the different one, that's the box library. Any questions about the interface, I think I have shown you all the interface of the program? Any question so far? [Quiet question from listener] can I have my own defined actions? Yes, you can create your own box, and then you can add this box to the box library. I will show you that in minutes. You can create your own boxes. Is it your question, to create your own box? Yes, just after. Do you have any other questions about the interface, about normal interface and how to use it? There is some button I forgot to mention. This one here, this is to control the volume directly in the robot. I'm trying to get my play sound on the robot. So you can control the sound directly here, but for you for simulated robot it's maybe not so useful, it's just for the real robot. You also have Battery level, here. This is the stiffness button, we have seen it. This is the Animation mode button, this one is very useful button to create animation. I'm also show you that, how to create some animation.

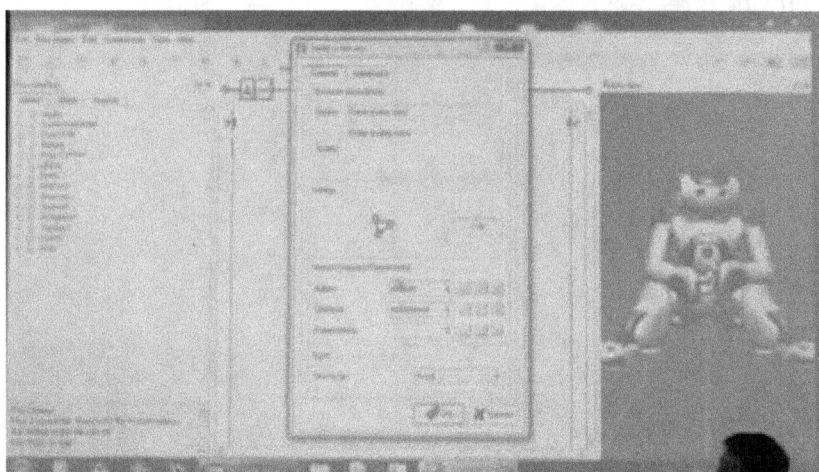

Any more questions about the interface? I will start to tell you how to create your own boxes. How to run your program. And then we will box library. And then I will finish simulator. To create your boxes, is very easy. You just right click in the diagram view and you choose "Add a new box". And you have this window appearing, and you can first give a name to your box, like for example "test".

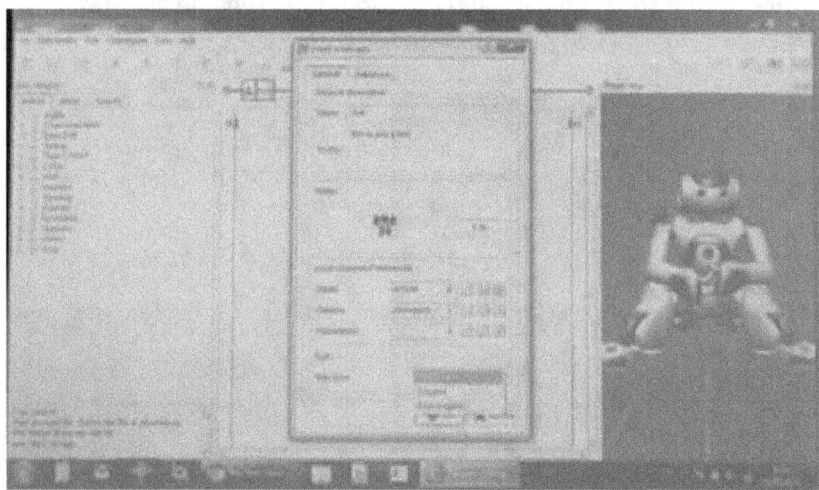

You can give some documentation also. You can change the icon; we provide some icons, which you can use, if you like. What is more important is this: box type, input, output and the parameters.

It is just define whole interface of your box. Let's start with Box type. By default we have three types: script, timeline, flow diagram. I chose script. Then, when I double click on a box, I get python script, because this just a box implemented some python script. If I choose...

so I can edit my box after creation, of course, and choose "Edit box". If I chose Flow diagram, then, when I double click on the box I enter into another level of flow diagram. So, that's the root level, and here is inside the box.

So you can make box inside box Flow diagram. You can have some levels across diagram. [Quiet question from listener] graphics path in the level... how do you go one path in the level, all the paths you above in the level... Yes. Actually, I told you, this is the main entry of your program; this is the exit of your program. As you entered to the box, you have also some input and output in the margin.

This is the input, main input of your box, and this is the output of your box. [Quiet question from listener]... Yes, depend how you link to boxes. Basically, this is just making some links through level. For example, I can drag and drop the Say box, link it to the beginning here, and link it the end here, and you will say text and then it will trigger the exit of my box, so then it will exit. But I can... just one box is enough, but I can do something like that, for example, I can change the language. First, I change the language, then I link to Say, then I exit. So, if I change the language to the French, for example...

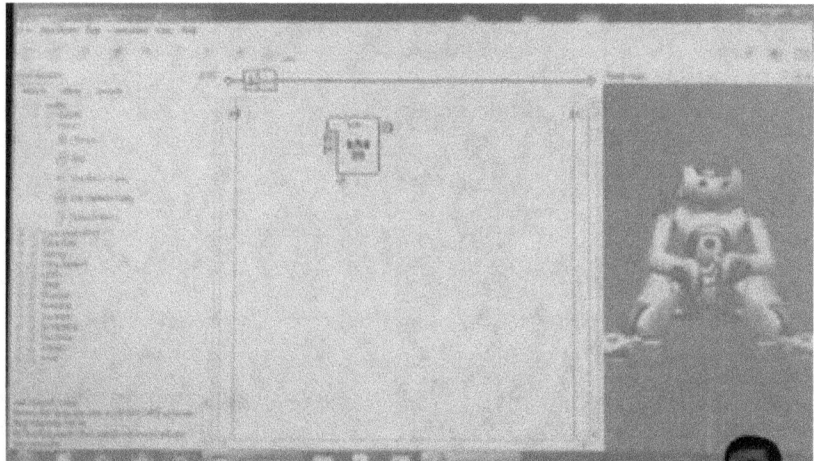

This is what is doing my box, you have group, two boxes into one. So if I link it, I can also link output if you want.

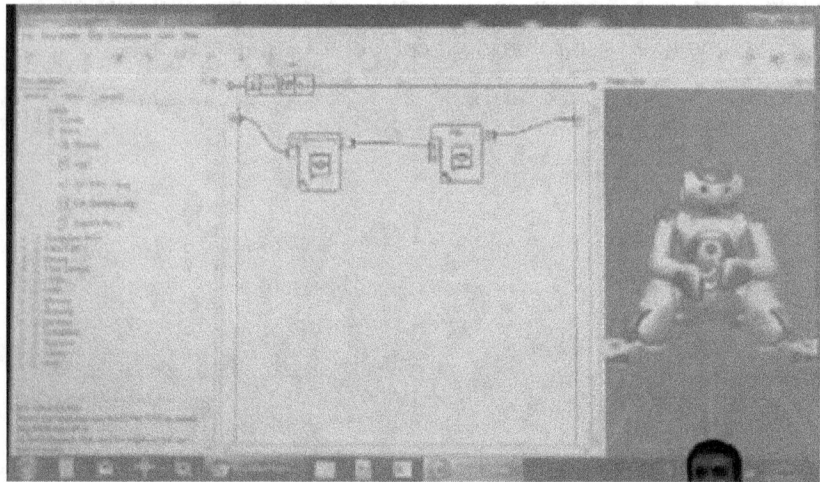

It just switched the language to Chinese and then say Hello. If you double click inside and you want to debug inside, you can also first play it here from inside.

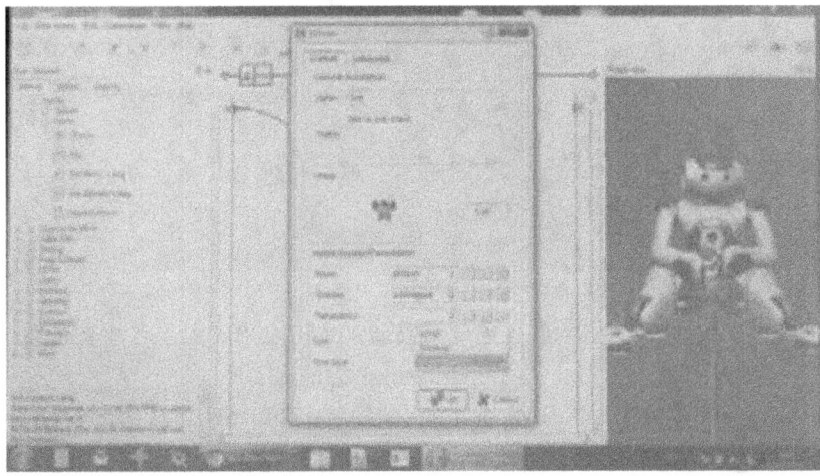

You can see how through it's going from inside the box. [Quiet question from listener] Can you define, like a switch case...? Yes, you can do a switch case with boxes, and actually we provide the switch case. I will show you that... different boxes we provide. Choregraphe is a visual programming environment, which means you can program it, just likes any programming but it's visually oriented. This is not the only visual programming environment in the world; there are other visual programming environments like that. You have Lab View, for example. Lab View is also visual programming environment with boxes and linking. It's the same logic. So, one box is a brick of behavior, it could be a big brick or a very small one. You can break it to smaller bricks inside. We just call those boxes. I'm switch back to English. So that's flow diagram type of box. Now, if I chose Timeline, when I double-click inside, I got flow diagram, but I also got the Timeline. The timeline is used to create animation. Basically, the time is going from here to there, you have numbers for them, frame numbers. And if you want to, I'll should now show you... It's a frame numbers, let me create box... name that box... timeline... this is the frame numbers and if you click the edit button, you can see the FPS, which is a number of frames per second. Here it's ten, which means frame number ten equal one second. Here is one second; frame number twenty is two seconds. Three seconds, four seconds, etc...

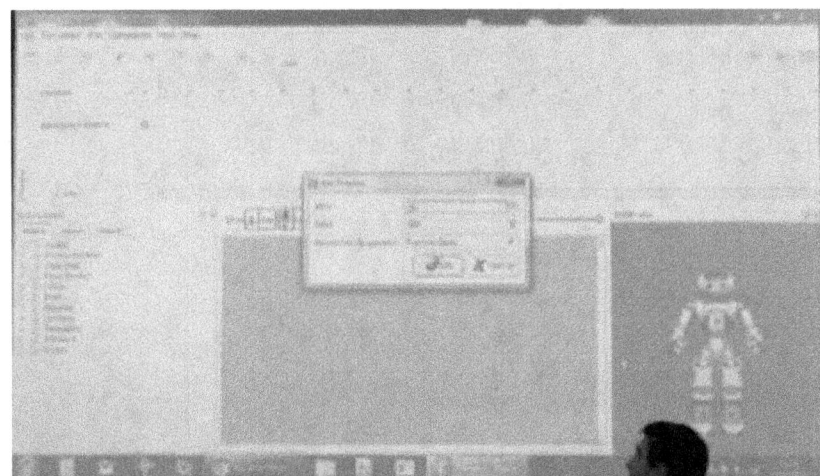

Basically, this is frame numbers. We can also have, in a timeline box you can also have behavior view. To have that you need to press the plus button on a behavior layer.

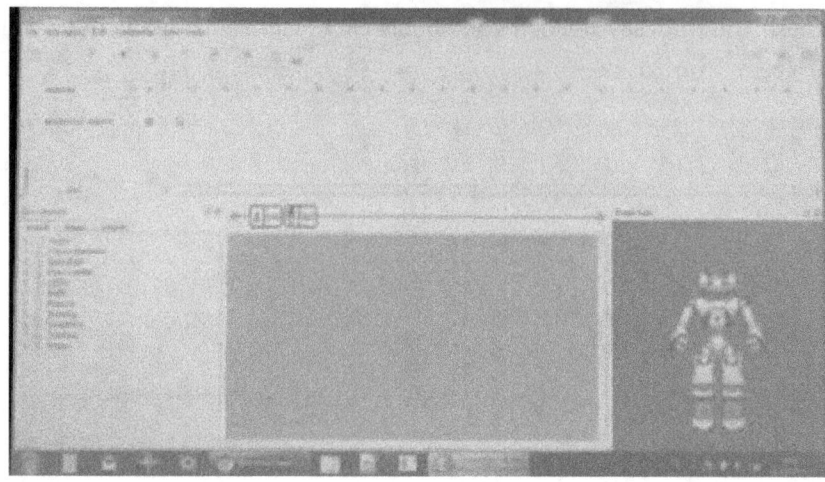

You can actually have several behavior layers, if you click several times on a plus button. And each one can have different behavior, different boxes inside.

For example, you can have Say here, and you can have another one linking other...

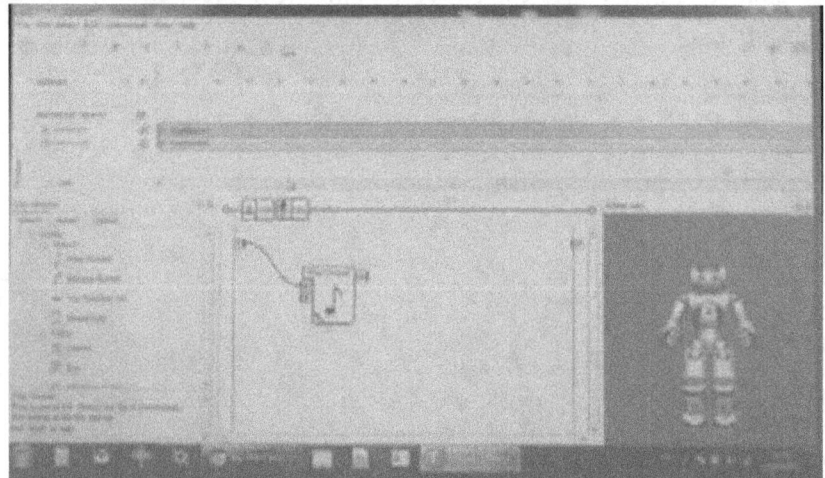

That means, what then the time is moving it will trigger the behavior. And I can move the key frames, so, for example, I can move this key frame here, this one. I just drag and drop the key frame, for example, like that. This is where you put the animation in the timeline. So, it's playing, what this box is playing, is playing the animation, and when the time reach the key frame then it start playing the behavior, to launch the behavior... it's like the key frame. You can also add more key frame, insert key frame on the same track, on the same behavior layer.

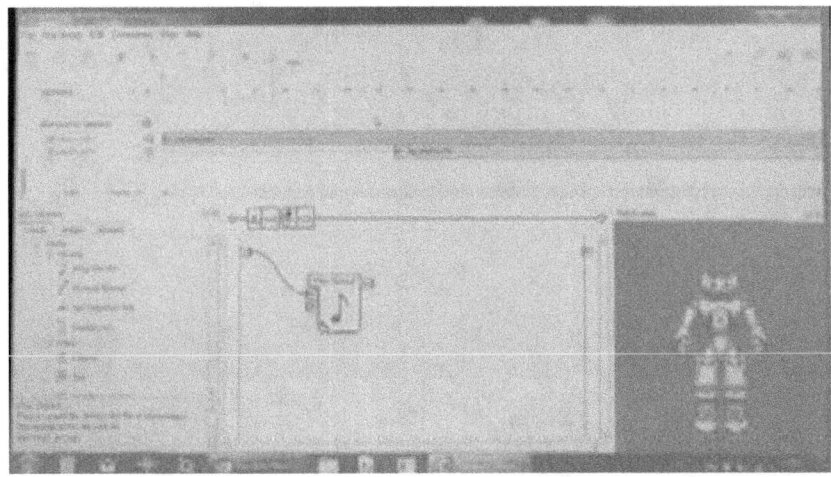

You just right click on it and choose Insert Key Frame. So, its mean when the time will reach its position it will start key frame thirty eight.

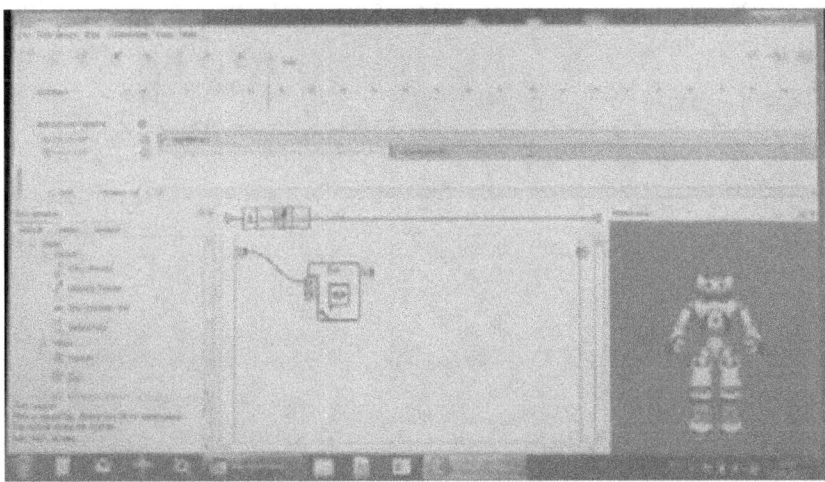

When the time will go until here, and then the time reach this one, it will stop... and stop... frame fifty one... And the animation, how to create the animation.

This is, usually, quite easy, I'm trying to do this time... It's quite easy with the real robot, but with the simulated robot I'm not sure. You can use the Animation Mode button here...

 I will briefly show you. You just click this button... robot makes the sound. And you can enable or disable the stiffness of the robot by touching some part of it. For the head you just touch the and you can put the stiffness, so maybe move head like that... it's not very important, the head. For the hand, for the arm, you touch the hand.

You can move the hand like that. If I release it, it fall... if you want to keep its position, I first touch the hand like that. And the stiffness is set. Basically, you can check the stiffness is set or not by checking the corner of the eyes... maybe you don't... If it's blue, that means that it's stiffness. Right eye blue mean stiffness of the right arm. If it's green, there is no stiffness. You can hold hand, becomes green, move it, release it, and becomes blue. Sometimes you lose the stiffness, because you can also voice control, voice command it... I say... "Left Arm"... "Left", yes, changes this... "Left Arm"... so, you see - it can switch off. Why I'm demonstrating you, in the same time I'm speaking and you just hear. We say "Arm" and reversal... stiffness sometimes... it's not very convenient... for demonstration. Honestly, the voice command is not very important, but what is more convenient is touching this. And for the arm... for the legs, you can press the button and the LED here indicates that there is stiffness. If it's green... no stiffness, if it's blue - stiffness.

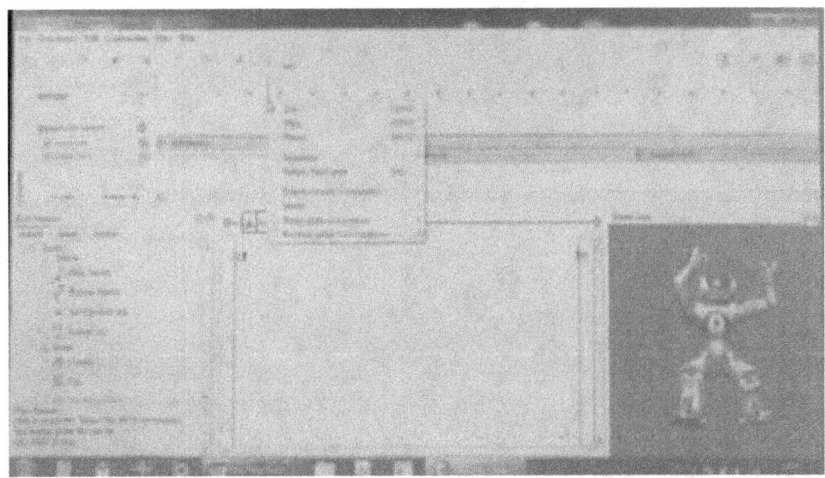

Basically, you can move like that... stiffness... perfect... of course, more like the robot... But, basically, to do that...... After that you right click on the timeline and choose Store Key Frame, that's why I didn't want to say anything.

You choose Store Key Frame and it will just record the positions, so, you can do it for many positions, and then, of course, then you try to do animation to play between all different positions.

OK, I will do very simple animation, to make you understand.

I forgot to mention, I may have broken my robot, we'll see. Forgot to tell you that, when you click on the timeline it actually go to that position.

That's why it's important to do that either on the ground or the place there you can catch it. Here was not the place I can catch it. As you can see, also there is shortcut, which is F8. If you want to store whole body. F8, F9, F10, F11. Basically, you just move your robot, just like I showed you and then you press F8, press F8 and it record several key frames, like that. Moment to do that.

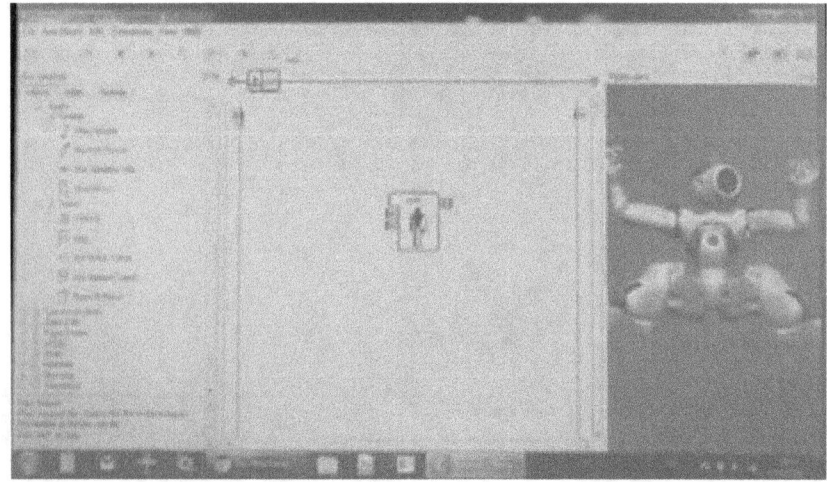

Record our number key frame. F8. And then record another one. OK, and then, when you have finished, you can exit Animation Mode and do the whole stiffness, and you can press the play button here, which replay your animation by blending between all the frame positions.

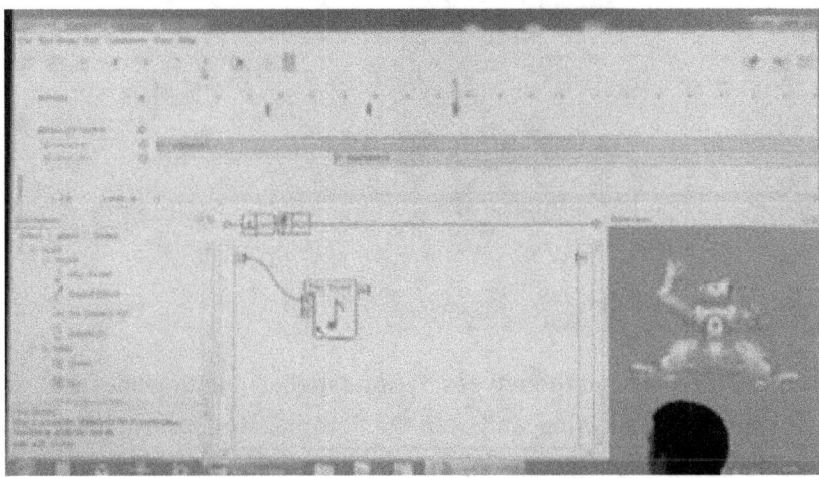

And that's just, this play button, is just for animation. But, of course, now if you want to play the whole box including the behavior you should, of course, play the whole... whole program.... delete key frame, don't need it. I just delete it. And I will try again, my behavior. And when it's finished, we can double click inside. When, you will see the red bar moving with the time. I'm launching the behavior. I'm sorry, you don't have the animation. But, basically, first key frame here is playing the song... here... so we start the song on the beginning. And here, on that timeline I just say Hello. I'm just say Hello a little bit later.

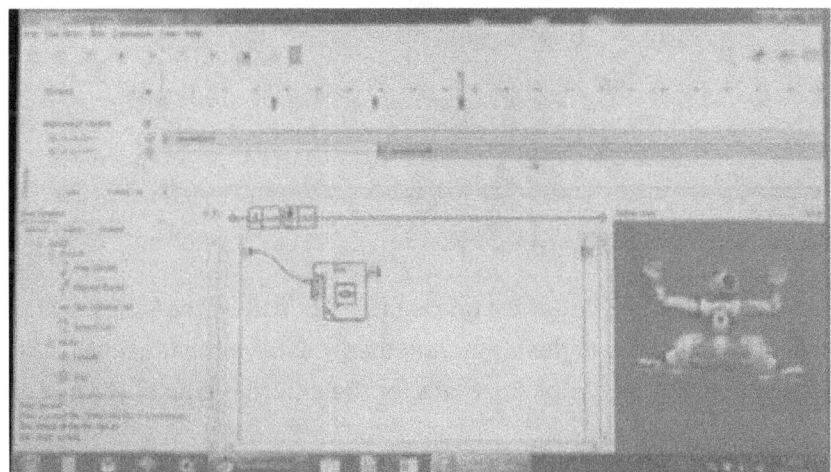

And I can, I just drag so I can synchronize my say Hello with my... key frame of my animation. This is the way to make it dance, to make some moves... say something or act or do something synchronized with the animation.

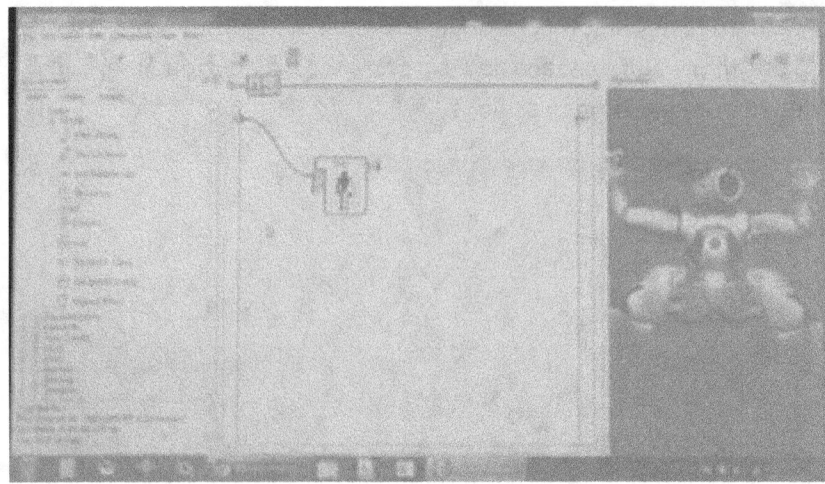

That's the timeline type of box. To summarize, we have three types. Script, timeline and timeline. Oh, Scripted, flow diagram and timeline. And even the timeline, and even the flow diagram box can have some script, you can edit the script. Just by right clicking on it. So, you can choose Edit box script and you also can edit the script.

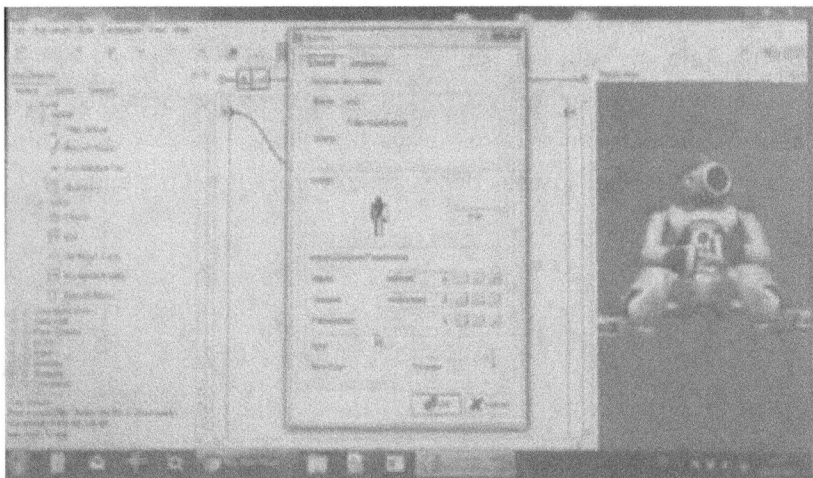

Basically, you have the different level of complexity. The easiest one is the script box on the script, then you have flow diagram, which contain flow diagram and the script. And then you have the timeline, which have animation timeline, flow diagram and the script. So, that's for the different type of the box.

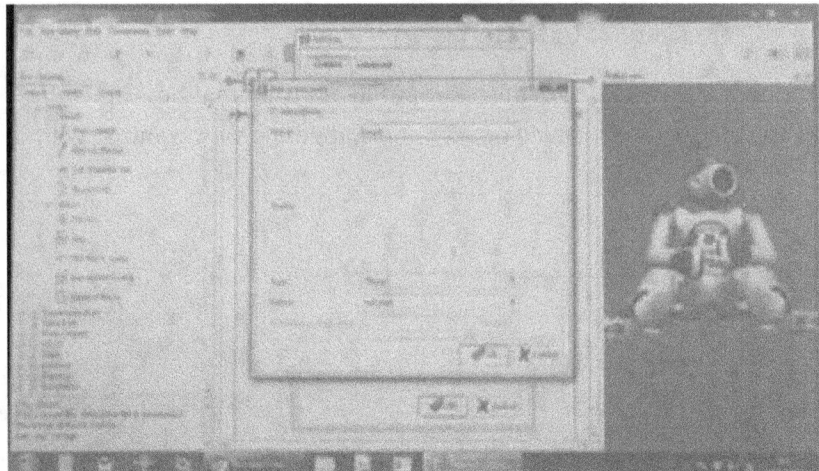

Any question before we switch to the next part? OK... Do you want to make a break? What time? It's five? So maybe at three thirty, forty? OK, now let's see the input and output. You can add some input and output; you just press the plus button.

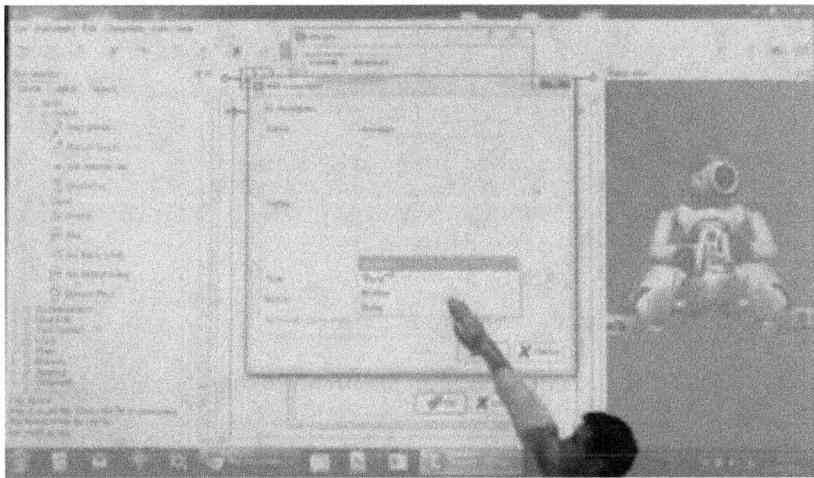

For example, here I just have an input. You have this window appearing. And, of course, you can type name for your input. For example, "Message". And you have different type for the input. You have four different types. By default is the Bang type, so the Bang type you can see as the event, it just triggers the input, just sending message to trigger.

Or you can have Number, or String or it can be Dynamic. And Dynamic means it can be any of the three others. Here, for example, I created String input. I will delete the default one also. Delete the other one. I will delete the output also. Instead, I will create a number input, like, don't know, eight. I will create Bang type. And I will create a Dynamic one. Here is my four inputs. And you can notice that we have different colors, depending on the type. The String are Blue, the Number are Yellow. Sorry, I switched them. The Number is Yellow. The Bang type is black and Dynamic type is Green. I can do the same thing for the output... Creating String... Number... Bang... Dynamic.

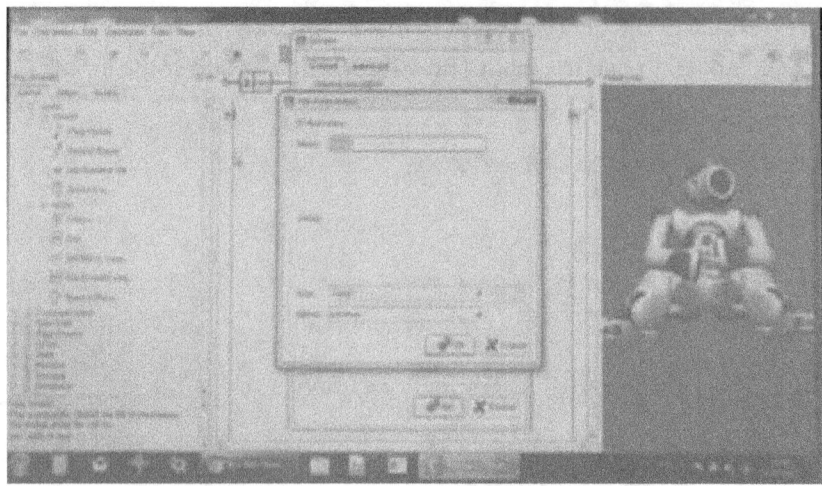

 OK, I have the same colors for the output. Now let's look how you can link them together. For that I'm just copy-paste my box. Copy, Copy-paste.

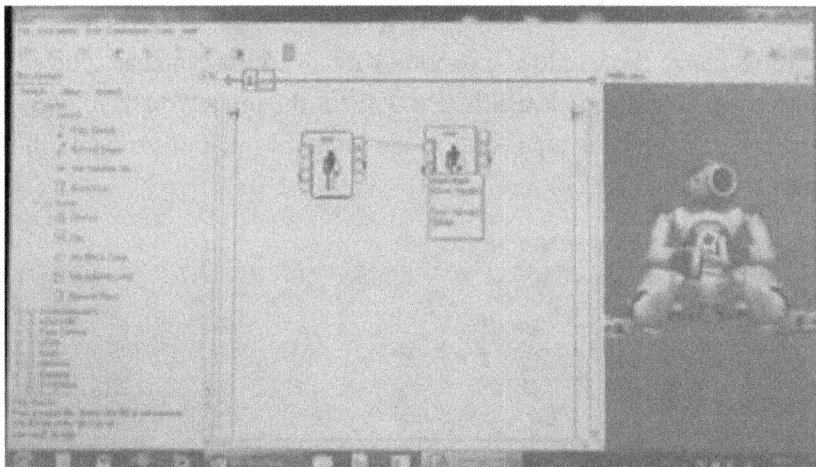

Of course, you can link string output to a string input, but you cannot link string to a number. Because a string can be... I don't know... "Hello, world" so you cannot convert to a number. You can link a string to a bang type. In that case, your value is just ignored, just don't care about value, you just want to trigger this input. You can also link string to anything, to dynamic type, because dynamic input can be a string.

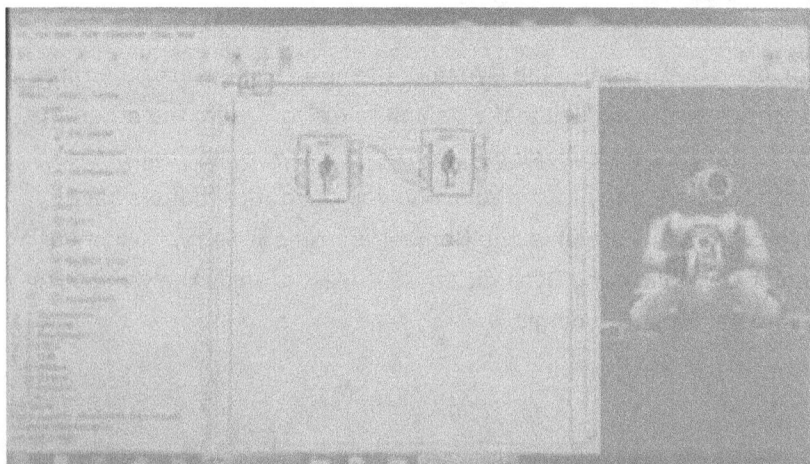

For the numbers, of course, you can link a number to number. You can link a number to a string.

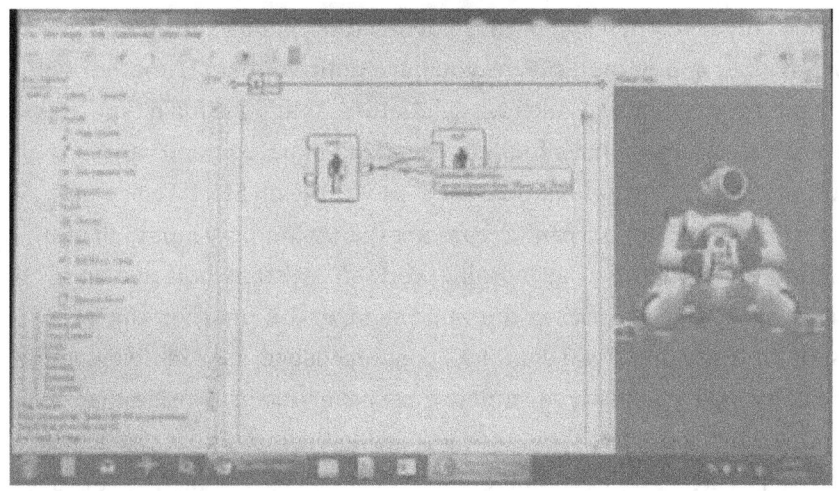

In that case a number is converted to a string. You can link a number to a bang, so that's the same as for string, we just ignore the value... we just define the trigger that is important. And you can link a number to dynamic input, because dynamic input can be number.

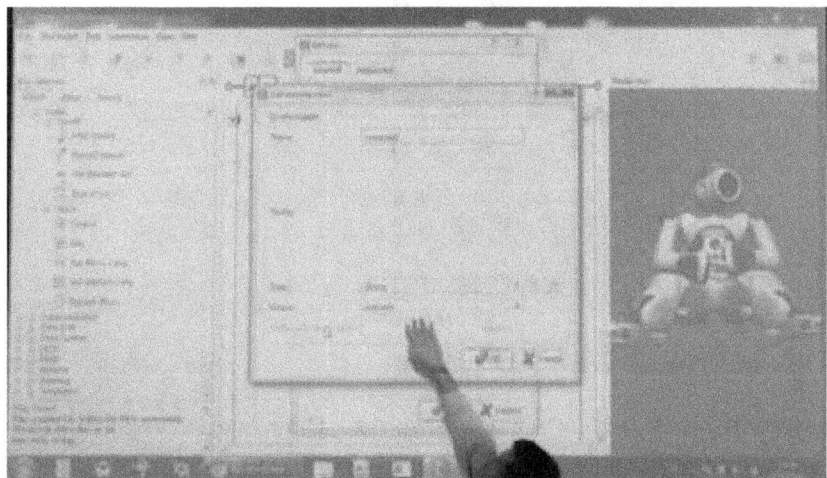

For the bang type, of course, you can link a bang type to another bang type. But you cannot link a bang type to a string or a number, because bang type - there is no value, no data. And the string input request a string. Number input is expecting a number. So you cannot directly link like that. But you can link your bang type to a dynamic type, because bang type can be dynamic. Oh, sorry, dynamic type can be bang type.

And, finally, you can link your dynamic output to anything, because dynamic output can be anything. So this is the link compatibilities between input and output. Maybe you have noticed, there is another type of... another kind of type... which mean for the input... Which we call Nature. That's the main type and we have another, which is Nature. For the nature we have four different kind of nature: on Event, on Start, on Stop and ALMemory Input. On Event is the default one. Let's look at the on Start, I will create message on Start. For the ends I will choose it on Stop. I can't create another ALMemory input... If you choose ALMemory Input, actually, this input, I told you this morning, you can register whole line on memory value. So, actually, this input will be triggered when the value change. You can give which key you want to link, you can also close it. For example, if you want to ...contain change, for example, we can move it like that. I told you, the color indicates the type, and now we have little icon inside the little box. The icon indicates the nature. On Events nature - there is nothing inside the box. These two are on Events. On Start nature of input - there is a little play icon. And on Stop - a little cross icon. What does it do? Quite simple, if you trigger the on Start input - it will start the box; it will load and run the box.

And if you trigger on Event input, nature of input, the box will not be started. Only if you have on Start nature of event the box will start to run. On Stop, the Red Cross, if you trigger this, it will stop the box and unload every single time.

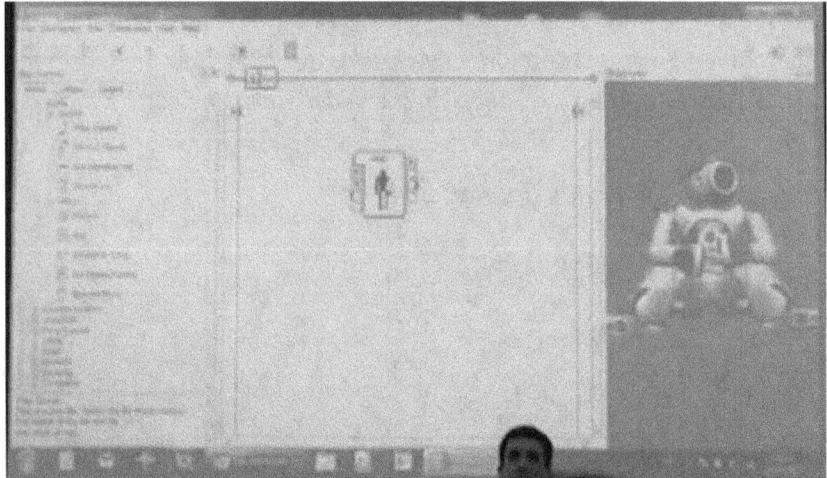

And, like I said, on Event doesn't start, doesn't stop the box. on Event input is just here to feed the box with new data, so if you want to send new data inside your box, but you don't want to disturb the running of your box, you just use on Event nature of input. If you look inside your box my box you can notice that you don't have the on Stop input. You have this one, this one and this one - is three, but you don't have the yellow one. Why? It's because you will stop the box, it's useless to start some new behavior, or something that you just stop the box.

Here, on this mode, you won't have on Start and on Event input. Plus, this little red one, which is ALMemory input, so this one is a bit shifted because it's that will come before Let's look at the nature of the output. For the nature of output we only have two: on Stopped and punctual, by default is punctual. I defined on Stopped and I have another icon appearing here. If this output is triggered, when it will stop the box.

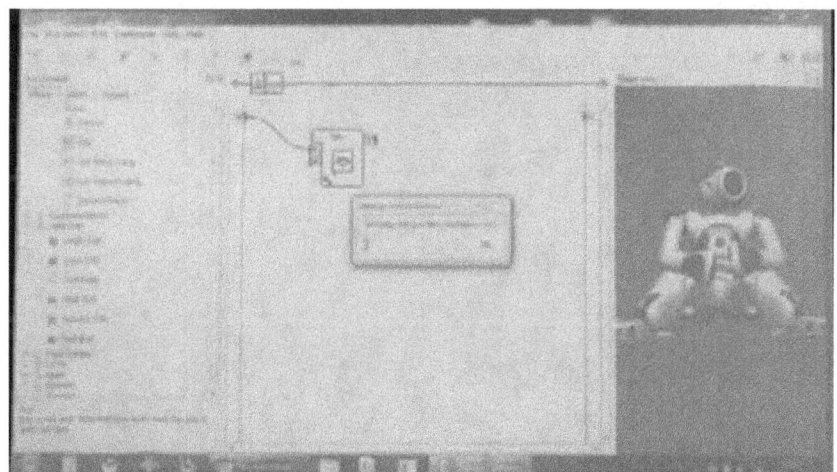

The on Stop nature of input is a way to stop the box from outside the box like that if I trigger this I cannot because it's a bang if I trigger this if I trigger this input I will stop the box from outside the box. And if I trigger this output, I will stop the box, but from inside the box Difficult? All the software engineers are. Basically, that's the difference between types of nature. [Quiet question from listener] When what about stop when it finishes? No, just stop, stop everything and everything, so you can be in the middle of animation, then animation will stop. [Quiet question from listener] Let's say you want about to finish you don't have to stop you finish anyway continue flow? No, if there is nothing that It depends on the kind of box, if it's a timeline type of box, this box will finish when the animation is here. For the flow diagram type of box or the script box - no, you have to trigger the output. You have to trigger on Stop output. [Quiet question from listener] Yes, just wait till that's why you don't if I can show you. If I just do that if I just do that my animation I should remove the if I just do that, the program is still running - that's why you have the red cross, because I didn't trigger the output here.

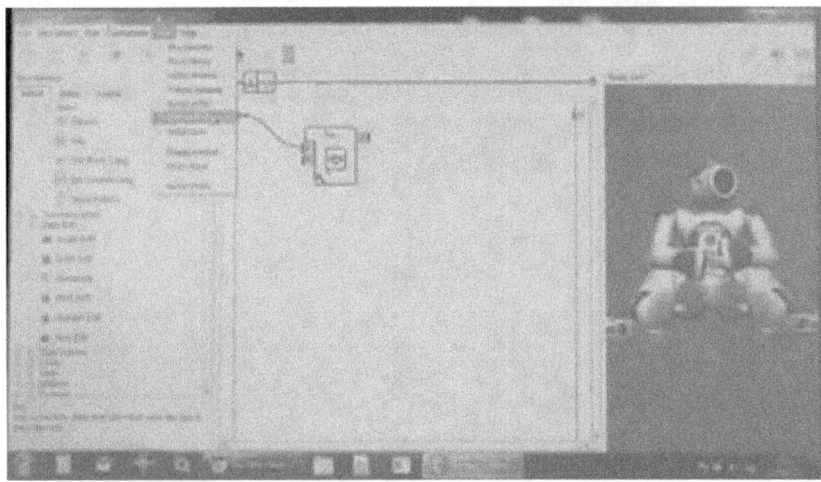

[Quiet question from listener] When you have started another behavior? No, no, you can if you play start, play again it will stop previous one, but it's Choregraphe that will kill, actually kill the current behavior. But now, if you link it like that stop it if you press play, it say "Hello" and then program trigger the exit stop, it will stop. I just demonstrated at the top level, but this is the same logic all the levels. Sometimes you want to trigger an output, sometimes you don't. And you want just to have something learning. [Quiet question from listener] Are killing this in low level you are actually will get activated stream stop? Stream you can use some form? Yes. Yes, exactly, yes, on the low level you just send a string out and stop your box. Maybe your box is computing your reasons and your reason is the string. It's the equivalent to return a string sometime.

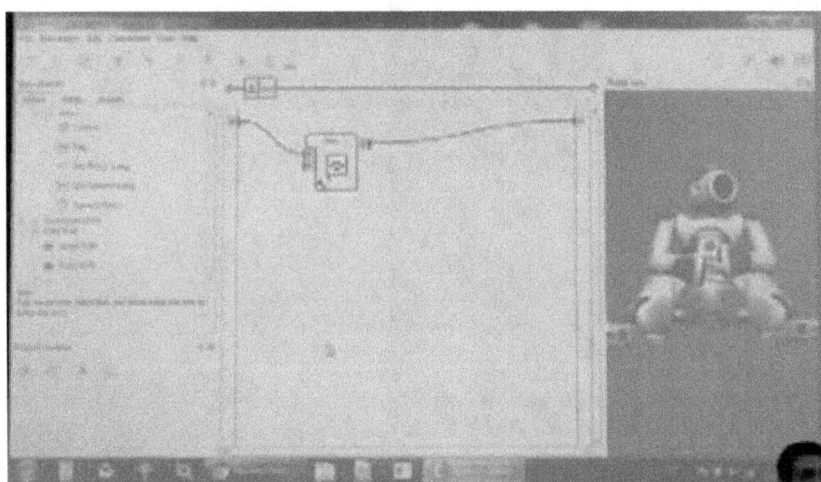

Yes, that's forth oh, I think I have deleted my box that's for the input and output. Now let's look at the parameters. You can also add some parameters to your box, just press the plus button.

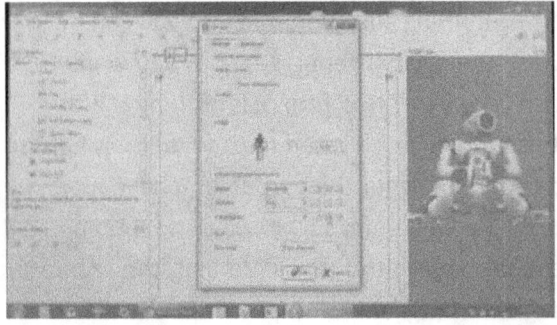
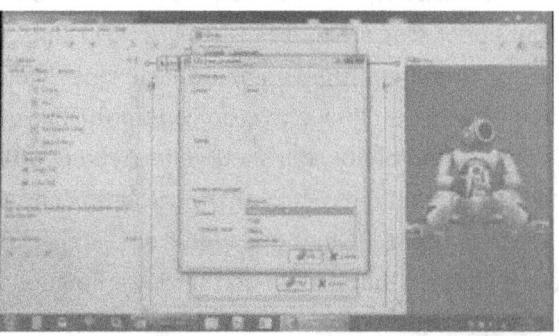

And you have different type here. Boolean, integer, float, string or file Boolean You can also specify the default value, for this parameter, if you want. Could be Check or uncheck, for this parameter will check.

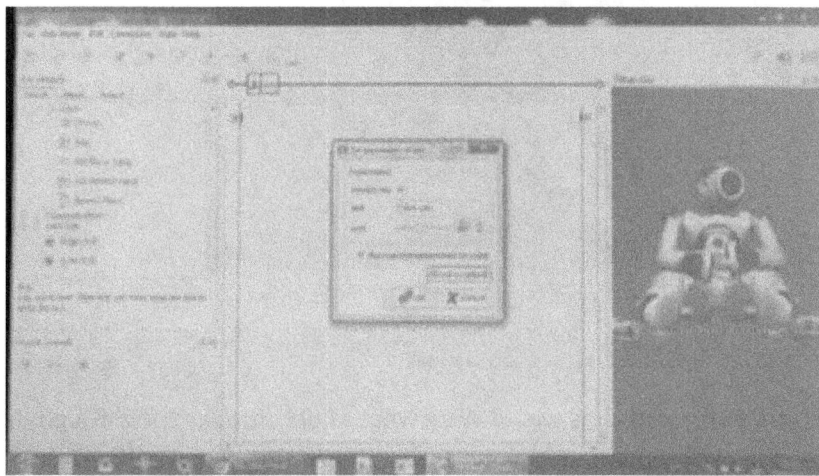

I create another one Text and you want to string. The default value when here. And you can also make multiple choice strings.

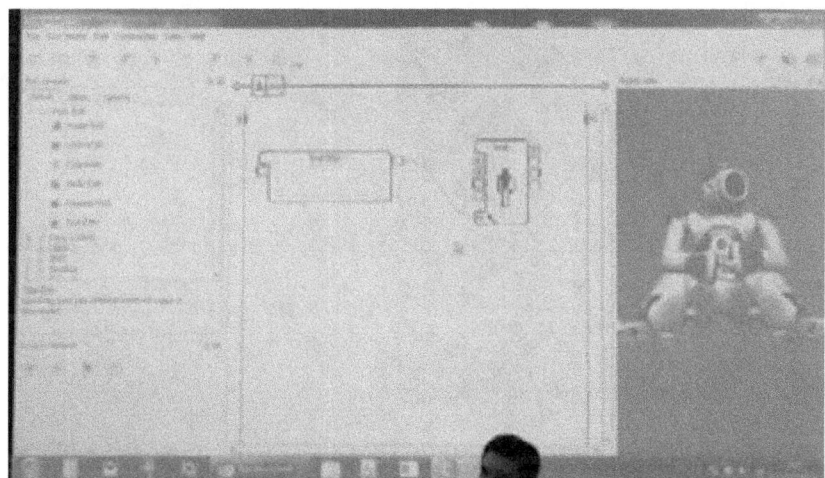

And if you choose number, for example Integer, you can specify number within minimum and maximum value. When you do that, you have a little wrench, in the corner here. If you click on it, you have your parameters which, of course, you can change to something else I don't know that and you can change the number from minimum to maximum value. And if you click "Reset to default" then your go back, your default value. Ok, so this, after this parameters can be used inside your box. [quiet question from listener] input Yes, just like input and you can even do even do real input, because you can actually link one output to the parameters, just like linking on the wrench and you release on the wrench and you can choose the compatible parameters. And for the one with text, and then you have a little palette here and here. So, you will tell me, what's the difference between input and parameters? And I just answer - this is the matter of design, how you design your box. And, usually, you create your parameters if you have a default value. If your box can work with a default value. And you put, you create an inputs if you really want the necessary input for your box to really work. It's just a matter of design. Plus, of course, you cannot launch a box just by linking on a parameter. If you want create a box, you really need an on Start nature of input, of course.

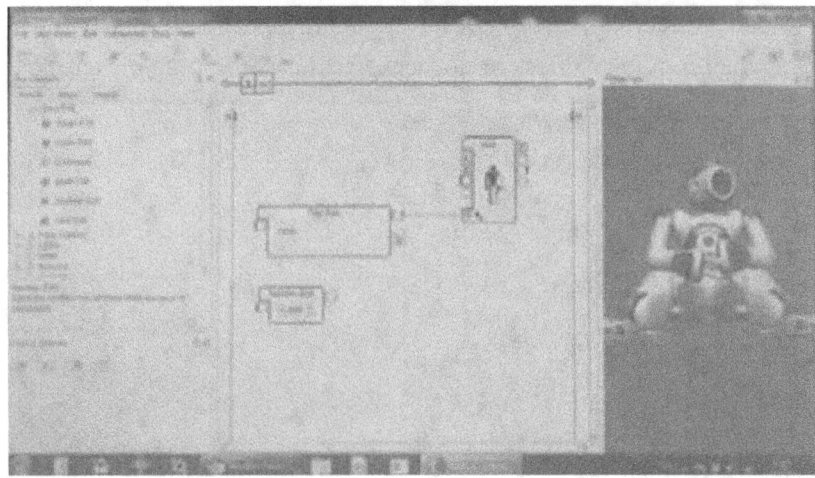

[Quiet question from listener] this strange volume box is what? Well, what's goes through the through the box lying just after, but I cannot certainly get things this; you can see this casting box. It actually cast Bang type to a String type. You enter with Bang type, which have no value, we give it value.

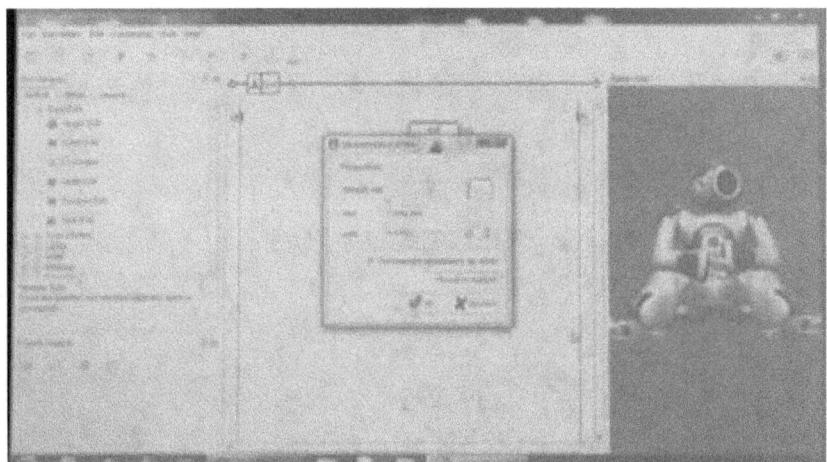

And then you have this string "hello", which will be send to the output, so it's a way to connect the black to a [1:30:36] and you also have a Number Edit box. I'm just doing the same thing, link to black one and it gets a number which you choose here. One more word also, about the parameters.

If you choose the Attach file, you can also specify default file name, for example. When you click on wrench, then you have this little icon and it will open the project content. If you want to have a parameters that load file from outside from your behaviors. Do you have any questions about input, output or parameters? No? I will show you something more about the script. I created new box, just default box. I will just create a new input with my name. Another one I will create with Eric name. This time I choose a number.

 If you just created box like that - it's a script box, so when I double click on it. You can see, that there is some functions created, some python functions created. And, basically, we create one function with name on Input underscore and the name of the inputs Eric and although it's a Bang type, so no parameters, an Eric is a number, number type input. As key parameters. This is python script, so for those of you who doesn't know python. That I know, very few people know python here, I may say something Python is a script language, interpreted language and there is three important things to know. The first thing to know is, it's tabulated language. Which means, there is no open bracket or close bracket for block of code? Instead we use the indentation, that's why there is little marks here, we don't see it here. One indentation defines new block. The second important thing to know. Is the "self" parameters. The "self" parameter here is the equivalent to "this" in C++. The reference of the inside of the object. Here, for example, we have a class. Class keyword just like in C++ it's the same keyword, defined class.

That's the name of the Class and that's the parent class So, the "self" is the equivalent to "this". And third thing to know, is that this scripting language some strongly typed variable [1:34:57]. The values are not strongly typed and its value can change. That's why there is no type for in this view. Can be string, number, can change Basically, for every input you have one function on Input and the name of the input, plus parameter or not, depending if there is Bang type or String type.

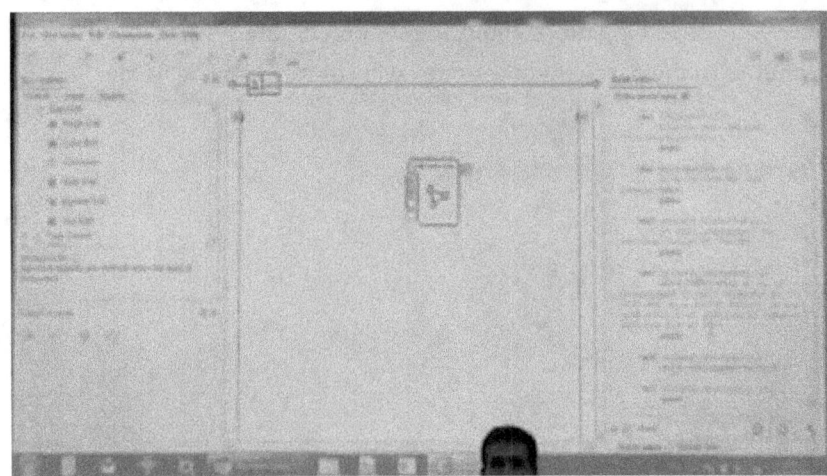

And for the output, this is the same, you also have one method for each output and the method name is exactly the same as the parameter. Here, for example, I have on Stop. That's the name of my output. It's called on Stop. That means, if you want to call this output for a script, you just type "self.onStop". For example, here about to see For example, if here I want to call the output, I will write something like "self.onStop". And if your output is a string or number, you can also get debug view. Like here, for example you could write something like that. If you change this output to to a String. It's quite easy, but just one thing - it's once you started to edit your script here, even if you edit the box by adding new input or output, it will not add the definition in your script. Because you started to modify the script, so, after that it won't touch your script. So, if you want to add an input after editing script, you can also add it past the definition in script. Any question? I think I will do the pause right now, because I think I have finished the explain you the programming intro, in box programming. That we can do have in Box Library. The entire box we have. And then I will finish the day by some NAOsim Give you fifteen minutes, something like that, the break?

3.5. Block library behavior

Now we will look into Box Library and after we will look at the NAOqi. Depending on the time, I don't know, we will not spend too many times on the box library. It organized into different categories.

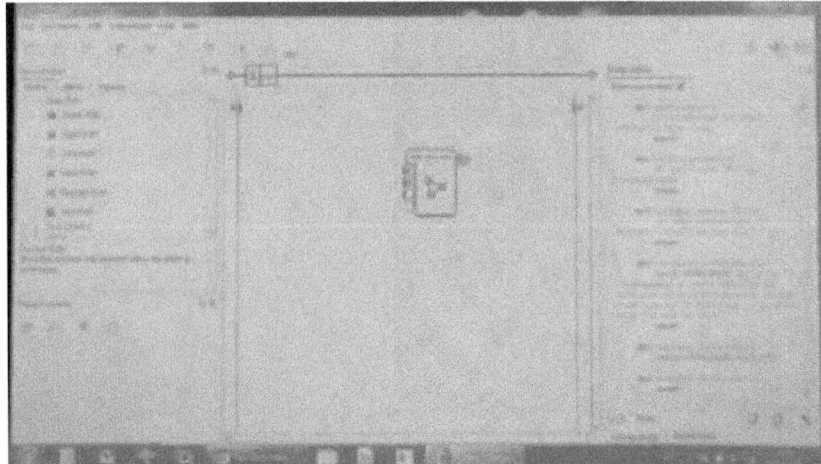

The first one is Audio - Audio box. We already have seen the Play Sound, which probably resisting [?] to work with the one pro [?] version.

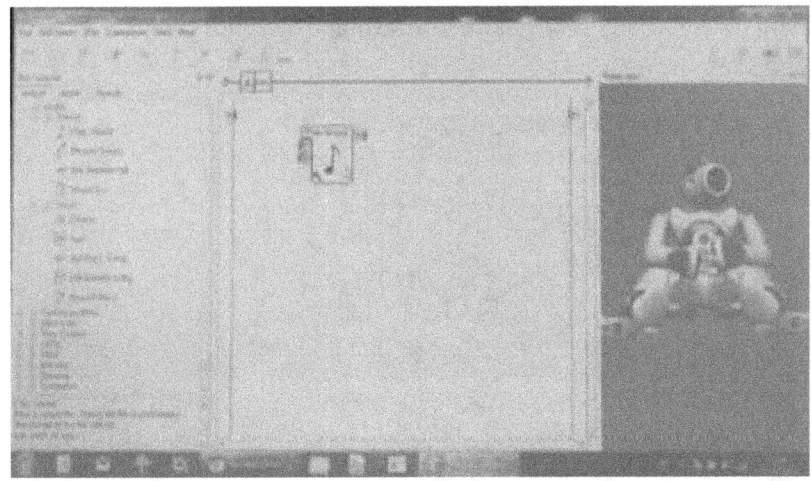

You also have little box to Record Sound. That's the filename, you can specify filename and what format: WAV or OGG and then, later, you can, I don't know - why you need to record this,   using speech to text. So, you recording, you can send your file to   speech to text, if you want. Then, we have a box to set the speaker volume. I think I can just have audio volume record time, very quick demo. I can actually replay the sound I just recorded. Just choose OGG, that's fine.

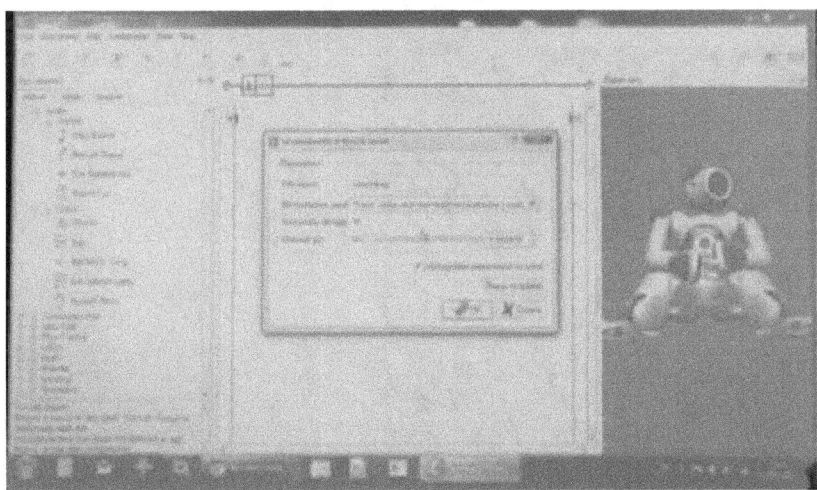

Then in the play sound I just type filename. "Hello, one, two, three - record my voice" [recorded] "Hello, one, two, three - record my voice" That was very simple, as you can see. You have one box to adjust speaker volume. Or you can adjust it here, of course, but with box you can do it inside your behavior.

We have a sound localization; I told you this morning, that's more interesting. The robot can hear where the sound is coming from four microphones.

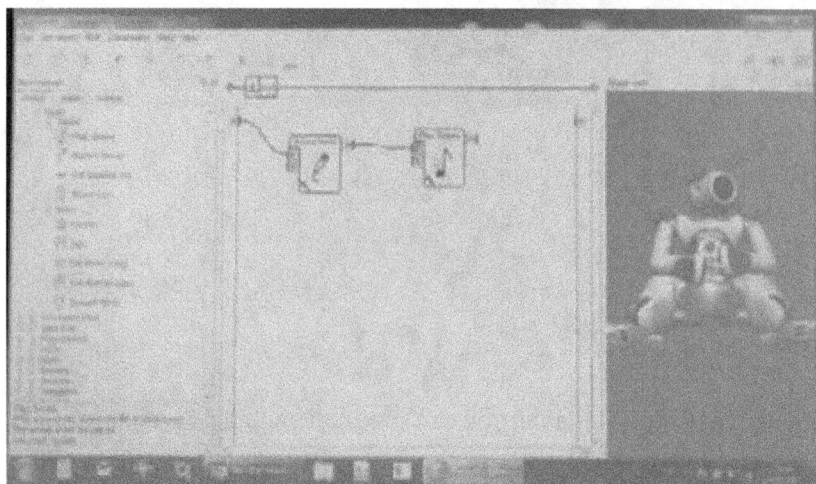

This box give you a value, actually, here you get a two value, two angles. Horizontal angle and the vertical angle. And it's related to the current head position. But, actually, we have another box in the trackers; you see it's called Sound Tracker.

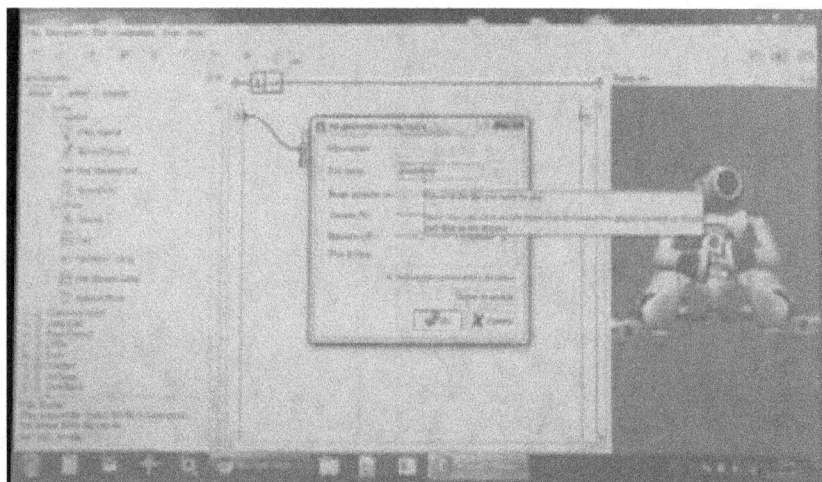

The Sound Tracker is using the sound localization box and it's linking results to HeadToTarget box to make the robot turn the head.

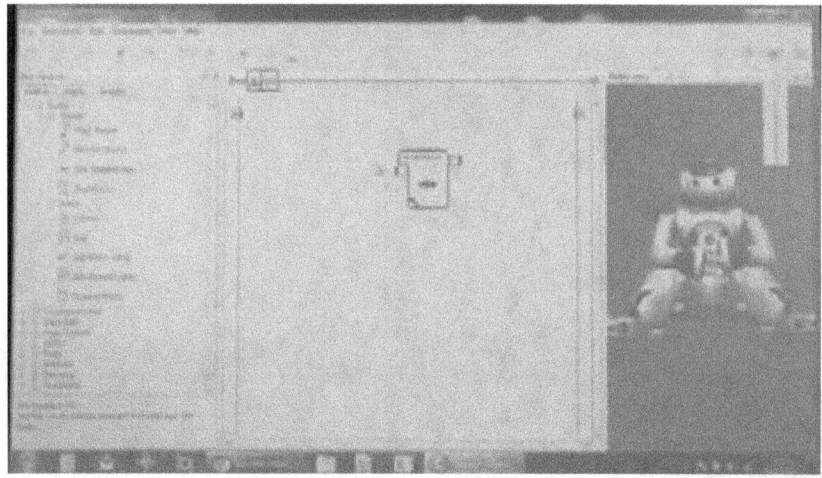

So, if we just try the stiffness, and go [Clapping and clicking] and so that's, like I said this morning, it can be used, if you want to have to attract your robot.

If you want robot to turn left, or even turn it and walk toward the people who is talking to him.

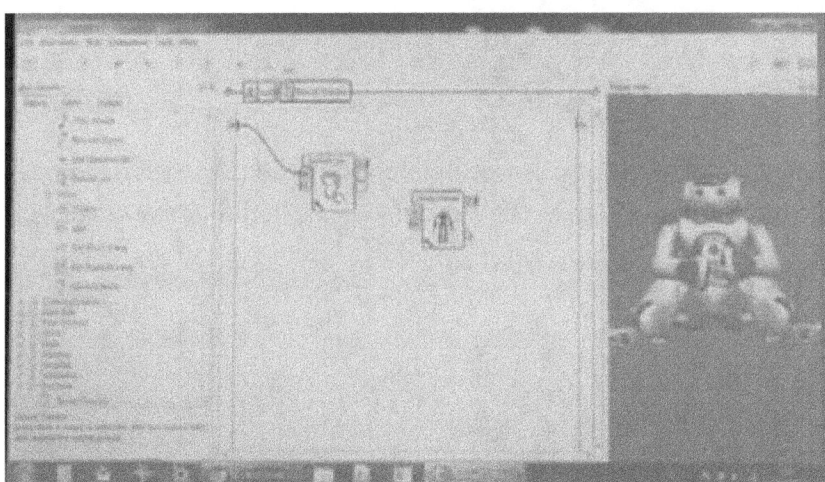

Or when he hears or not. That's surprise. Usually, when I show this demo I have many people shouting "Hey!", "Hey!", "Here!", "Here!" - You are very quiet.

[Quiet question from listener] How he knows where the sound coming? The robot has four microphones. Release the stiffness. Around the head there is four microphones and depending on the volume of every sound maybe the sound coming from here being higher, than the one coming from this here. So it computes what is the angle depending on the sound from this four microphone [quiet question from listener] Height Yes, yes, because you have also here is higher than this one. And this one higher than this one.

00:05:00

 Three not in a plane, but really spaced. [Quiet question from listener] Yes, I don't know details because it's some other engineer we have basically, it's just very sound for four microphones. You have also Say Box, like I showed you the same box on another level.

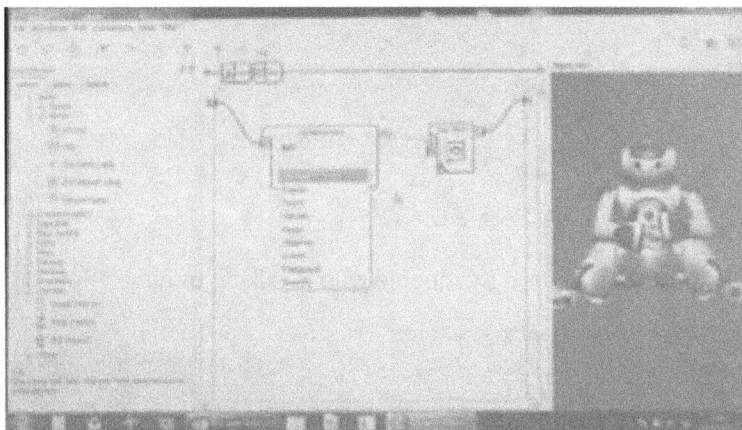

This flow diagram box where you can type the Text. You can type different translation of the same text, but you need to switch the language on the robot - to hear the text pronounced in the correct language.

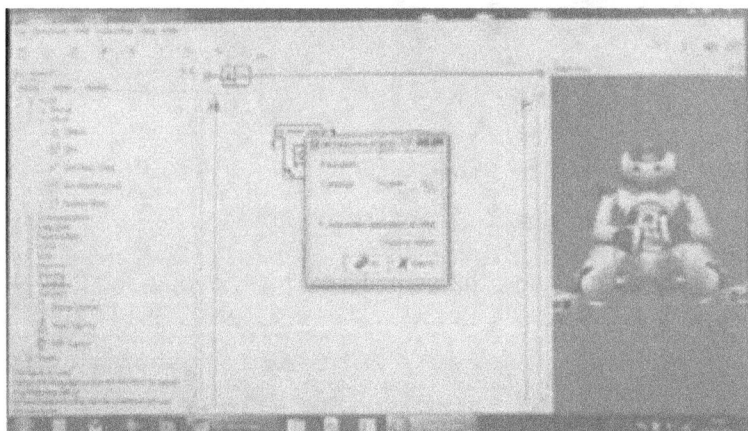

And to do that, is the Set speech language box, you can use this box to switch the language on the robot. You can choose which language here.

Like I said, you need to have installed the language package. There is little limitation - you cannot have more than two language installed on the robot.

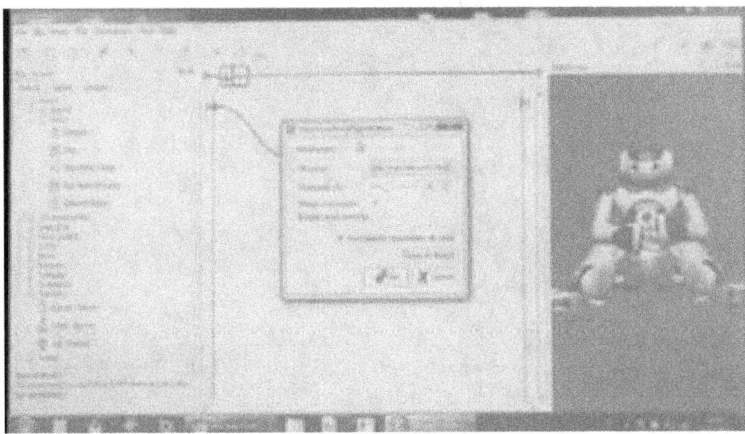

Not we, because I'm understand that you still can only work two robots to languages on robot. So, license issue. Only a license prohibits this. You can also set recognition language. You can set speech and recognition - different languages; you can talk to you in French and vice-versa. And you have a Speech Recognition box.

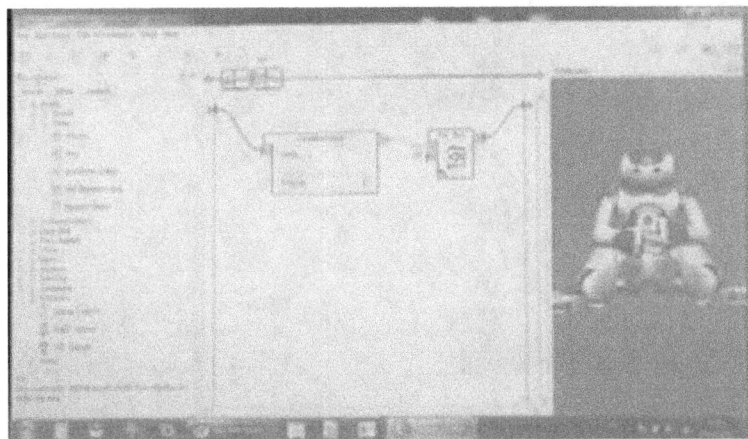

This one, you can specify list of keywords you want to recognize. By default Yes/No, but here you go [murmuring] and I can ask him to say what have just recognized, for this I will take the same text. So, this box, which take string to say I copied, pasted here, don't need this one anymore. So, I can link output word recognized to Say Text. I should also stop the speech recognition once we have recognized one word, it's a bit like for the vision recognition, that's the same story, because if it don't stop, the robot say the word and then recognize what just said and say it again, so we have to turn off.

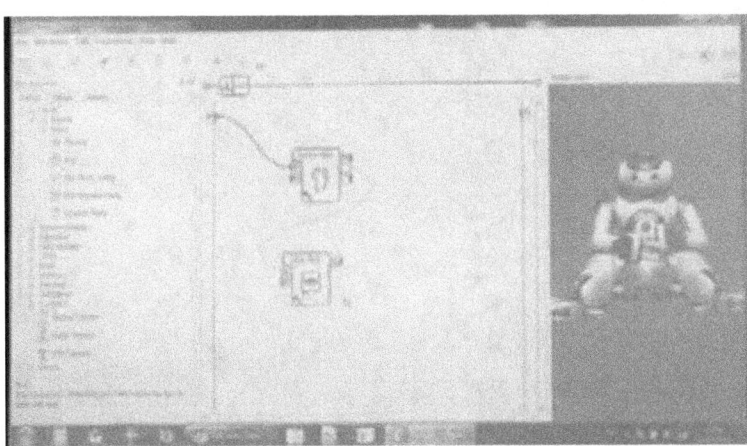

And I can restart the speech recognition after the same text.

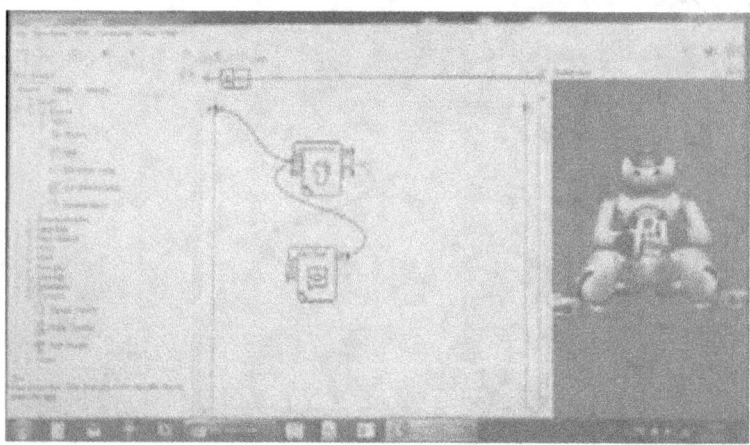

"Chair" - "Chair" "Television" - "Chair" "Television" - [silence] "Table" - [silence] "Table" - "Television" He is not happy with this - "Table" "Bed" - "Bed" "Table" - "Television" But, basically, the speech recognition is working quite well.

I've shown you many times this morning the application where I talk to the robot. Yes, you can use this one. Basically, when you wake behavior you can use you can enter into so part of your behavior where robot is listening for self-answers. The more answers you took, the more errors you can get: "table", "television" - they are a bit similar, so, of course, the less answers you can have - the less errors you have. We also have the Choice box, which is more complex box, because this one can enter multiple cases, multiple situations. If you enter into the next level, you can actually see under is two main boxes. First, the question,

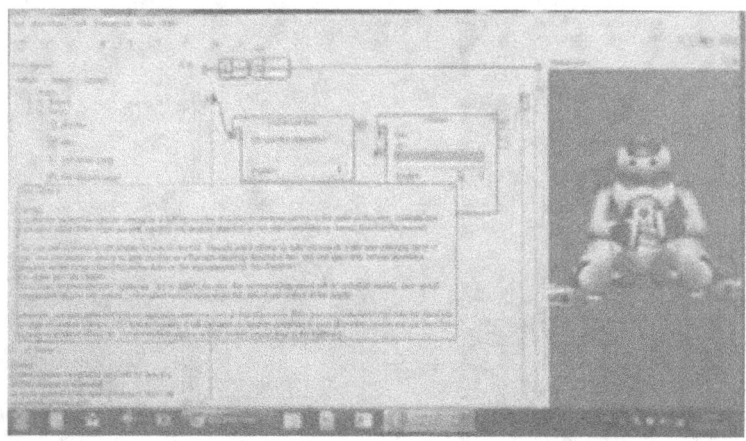

And then - different choice, different possible answer for this question. You can add more you can change your question, you can add more answer, like if you do that you should also link to your output here.

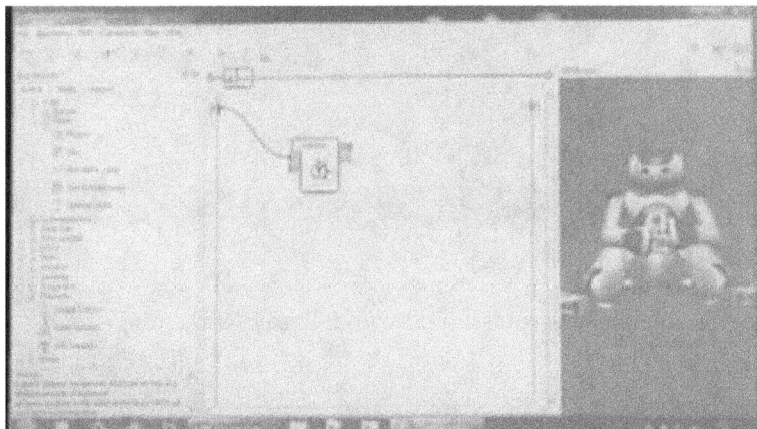

And then, depending on which answer it have recognized you will give your answer here. From here, you can add the switch case; this is where I answer your question - the switch case. Because we actually, I didn't tell you, we actually have the switch case - it's in the flow control. We have the switch case here.

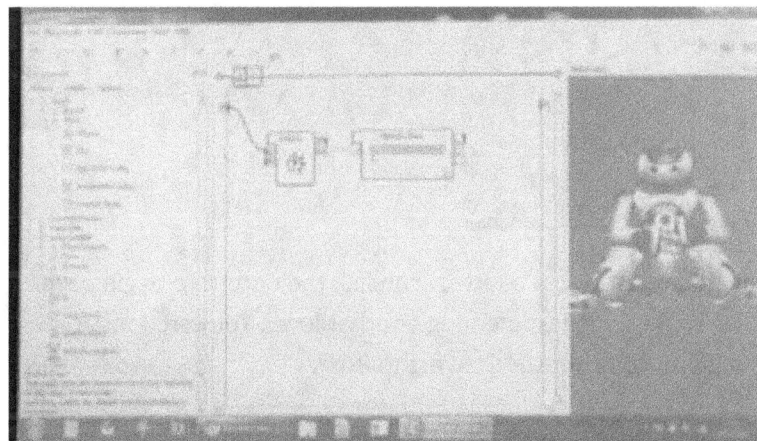

You just link the word recognized to switch case and of course you should try different possible case, which are Yes, No and then, from here, this numerous switch case, which means you can trigger different action for the robot. For example, the robot could use animation, like, for head.

If you say Yes, and say something, maybe you say No.

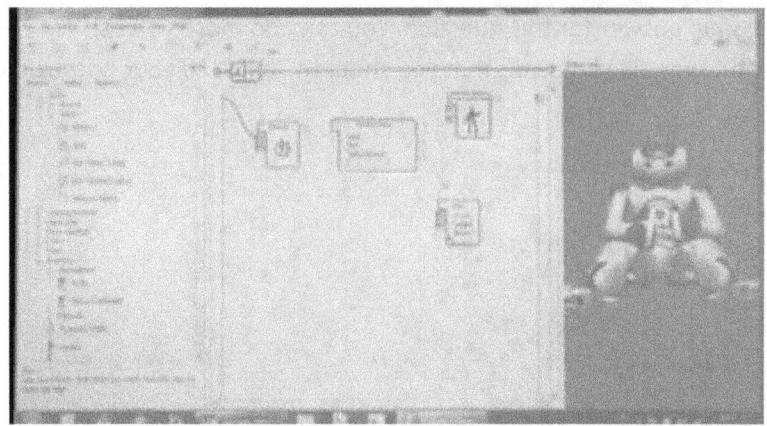

For example, if you say "are you nervous" you tell me "are you stupid? » And if I say "No", you just say "Hello".

If I want to look this behavior and ask again the question, of course, I can link the output to begin of the choice, but it can became a bit messy, like that. So, I suggest creating another level. You can do that - you select all your boxes, you right click on it and you choose "Convert to Box".

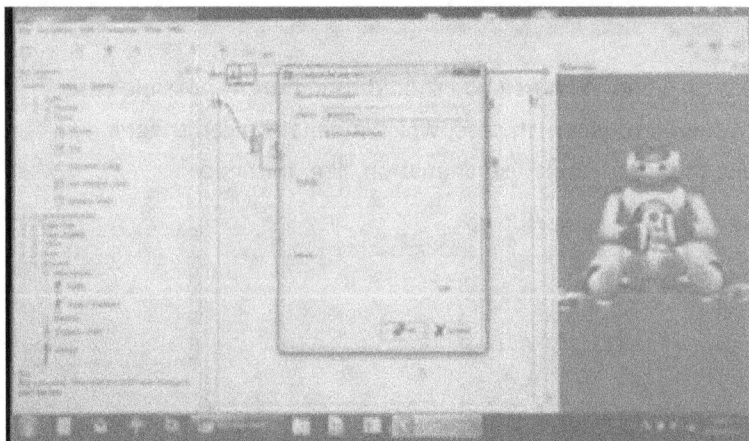

For name - Call your questions, you need to type anything specify image. Then you have your "questions" a new question box.

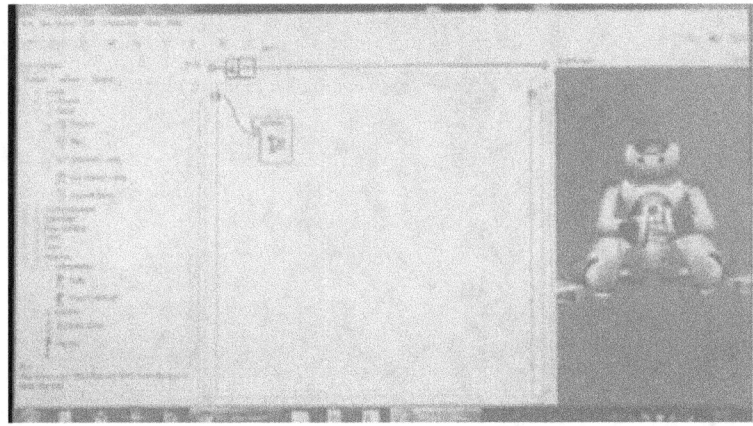

If I enter inside, I can highlight my own choice boxes. I can add some outputs. For example, I can add exit output.

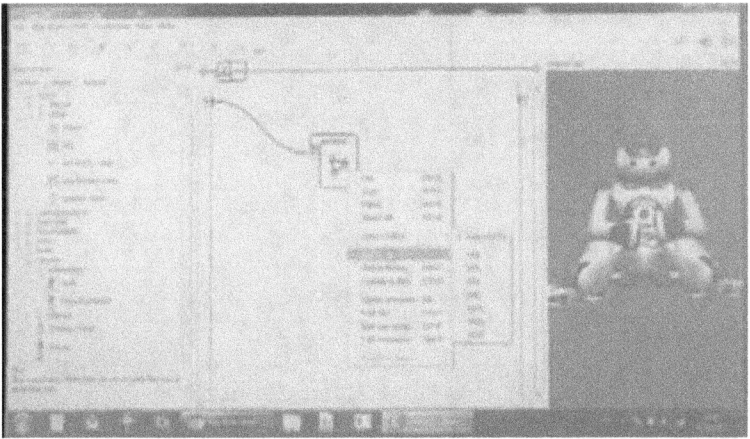

And I can add finish output. If I exit, I just exit the full out, if I finish, I start again the question. And here, for example, I can, after reacting to my answer, I can just output finish.

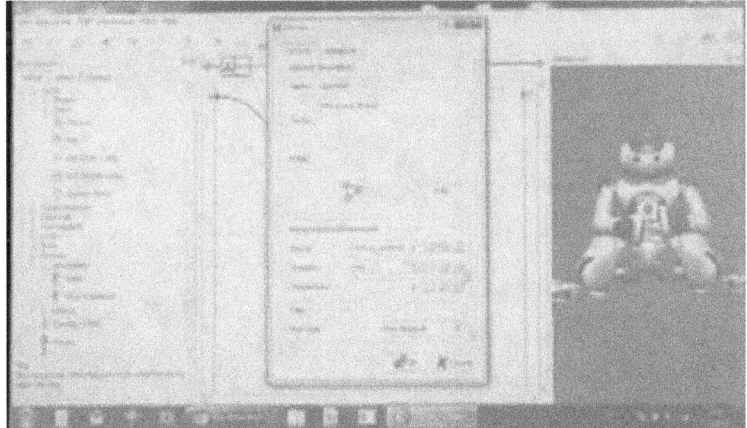

And here, if I didn't recognize any word or if I decide to exit to skip the question and then skip exit. Let's try that on a robot. "Do you like chocolate?" "Yes" -"Yes".

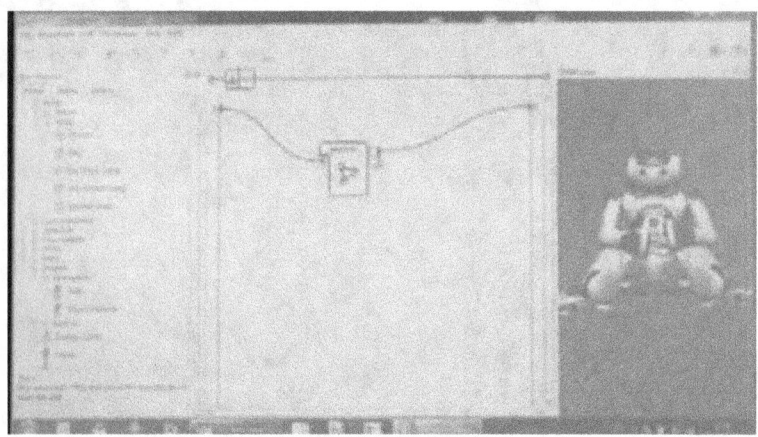

Oh, I didn't put the stiffness, so that's why it didn't wave the hands. "Do you like chocolate?" "Yes" - "Yes".

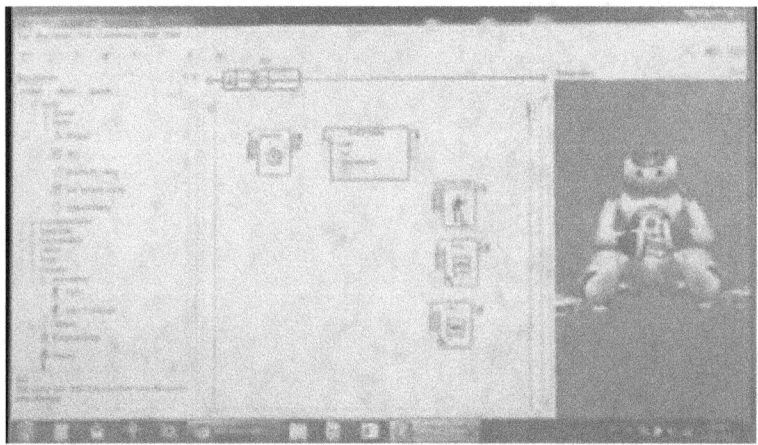

"Do you like chocolate?" "No" -"Your answer can be Yes, No you can ask me help you can also use like that"

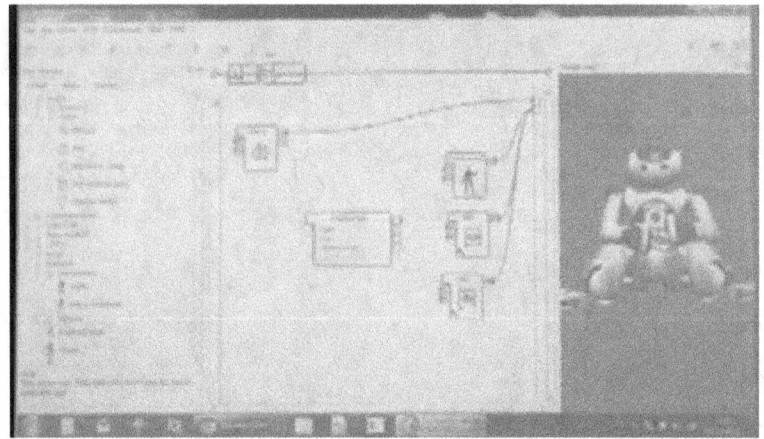

So, it propose to some help "Do you like chocolate?" -"Repeat" "Do you like chocolate?" - "No" - "No. Hello" "Do you like chocolate?" -"Who knows" "Who knows"

00:15:00

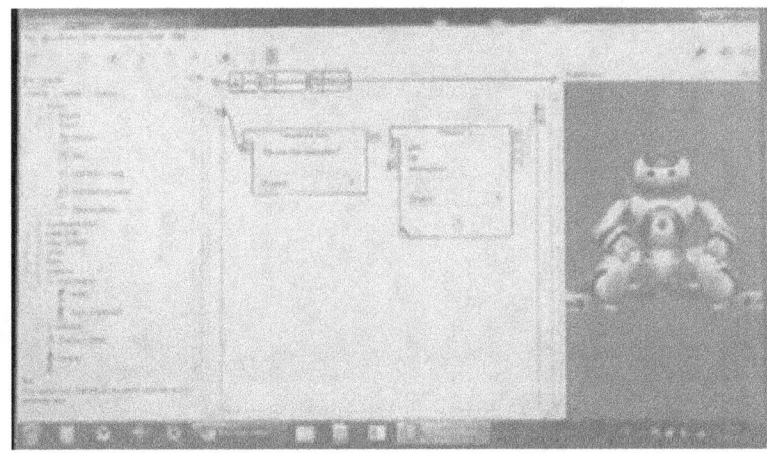

It should say I'm stupid. It didn't say I'm stupid. Because I didn't link this one, here. And now the program is locked. The bug was - I didn't link this output "who knows" to the exit, so the program is now stuck. Easy fix to link it. You can stop it. "Do you like chocolate?" -"Who knows" - "No, hello" "Do you like chocolate?" -"Who knows" - "Who knows, you stupid" "Do you like chocolate?"

So, you can also use tactile sensor "Yes, that's it" You can choose it, you can draw it backward and then press in the middle to choose it.

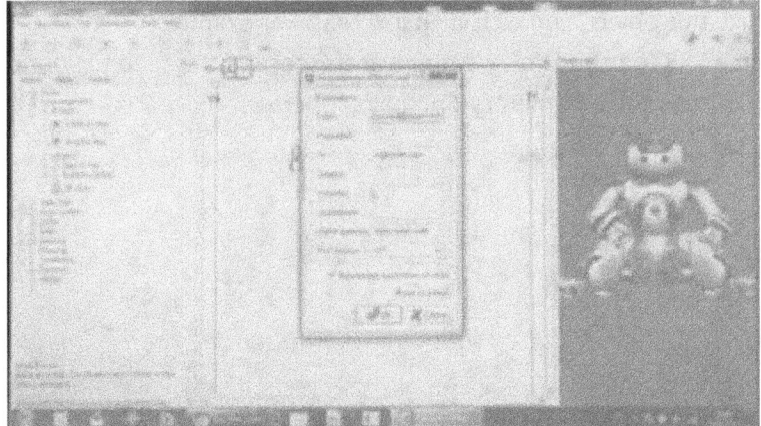

So, you can see, the choice box is a very complete box, you can evaluate, evaluating box. "Do you like chocolate?" Which can all the case so if you want to have interaction with people it's very easy, you just use the choice box and you can have ask any question to the user. OK, that's for Audio.

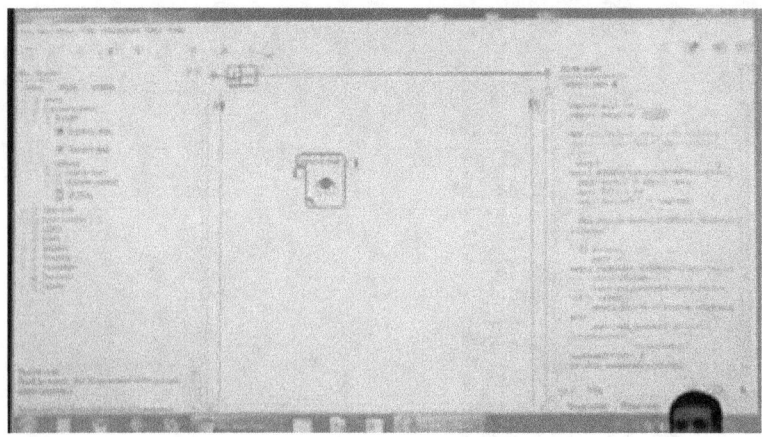

About communication, oh my god, you will be disappointed. We don't even have Wi-Fi box but we will have soon we have two email boxes, which can send or receive email.

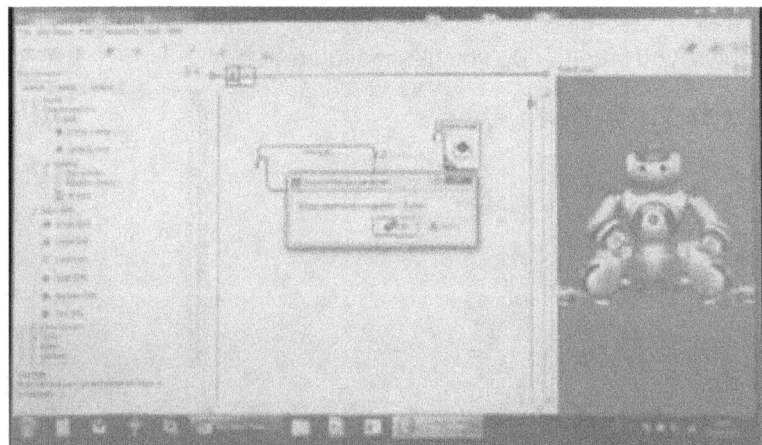

So, Send Email. If you click on the parameters, you can see email from, to, the SMTP address, the Port, etc. So, you can send email. If you actually, if you look inside the box, it's not magical, it's just some python script and it use smtplib, email libraries from python, so the box is almost have no code correct function of the python library, it's very-very easy, very lazy work. Also remember, you can remember that you can make this box more dynamic. Because you can link to parameters, the wrench. So, if you want to have something more dynamic, like this object, you can link it like that. For example, I don't know, you get all the content you get, some RSS-feed and you want to transfer resend this feed to your friend, so you can get the feed and then link it to the parameters. So, that's for email.

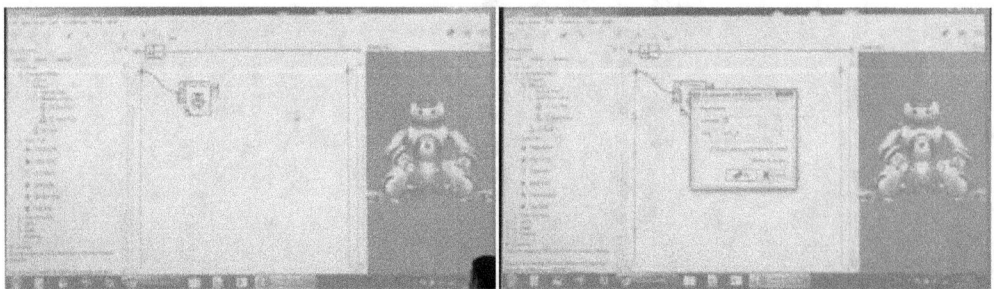

We have infra-red communication with the remote control. This morning everyone you got remote control work and then uses it for example, infra-red Get Key, so, if you start this box and here you will get the key that you received.

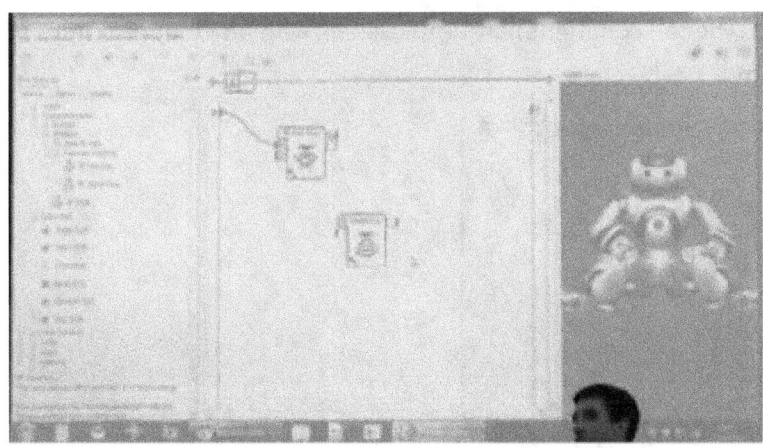

And NAO can also send the key if you want to control your TV. In parameters you choose which is what the name of the remote is. This morning I called this remote Epsilon, so I first type Epsilon, and then you choose the key you want to send. And if you trigger this, then it sends the key in the environment and if NAO is looking at the projector, then NAO will control the projector. We have also some very basic Mouse To [?] makes some infra-red communication between if you know about that policy you are very network guy, and IT guy. Forget about this box, it's just some simple example mouse. That's for communication. Data Edit.

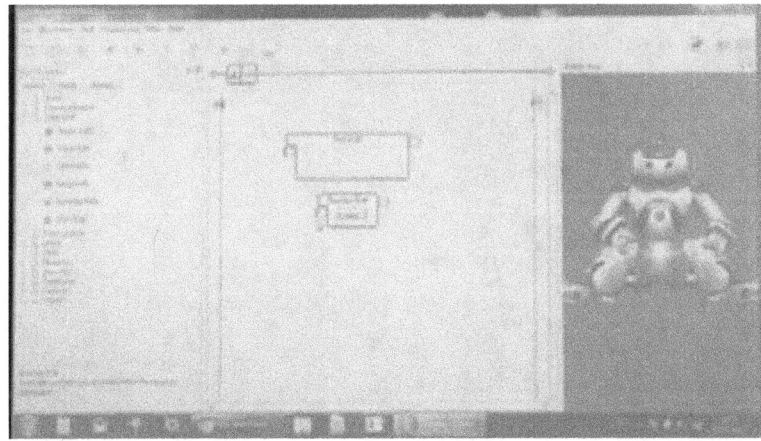

I already mentioned, that is Data Edit with the Text Edit, with the Number Edit. You can to a number or to a string. And you can also have a combined interaction [?] between a bang to dynamic type.

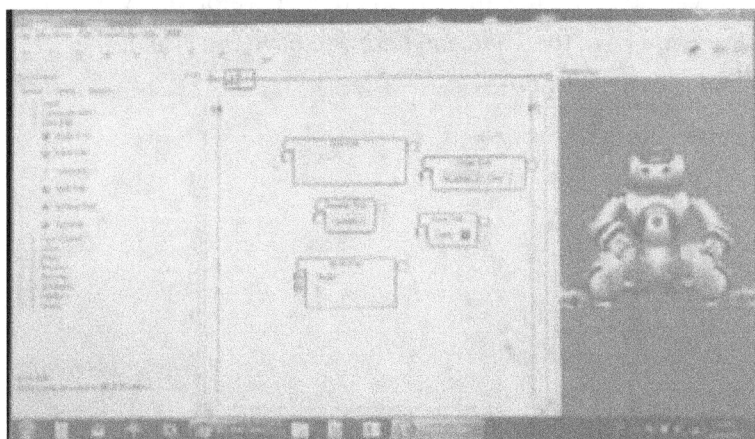

You also have some Angle Edit or Color Edit, which is basically the same, with just nice color-picker and it converts the color to a number. It's nothing but color, so it's convenient. And you have a Comment, if you want to comment your behavior just type the comment here. This is a box with no input or output. Flow control - that's more interesting.

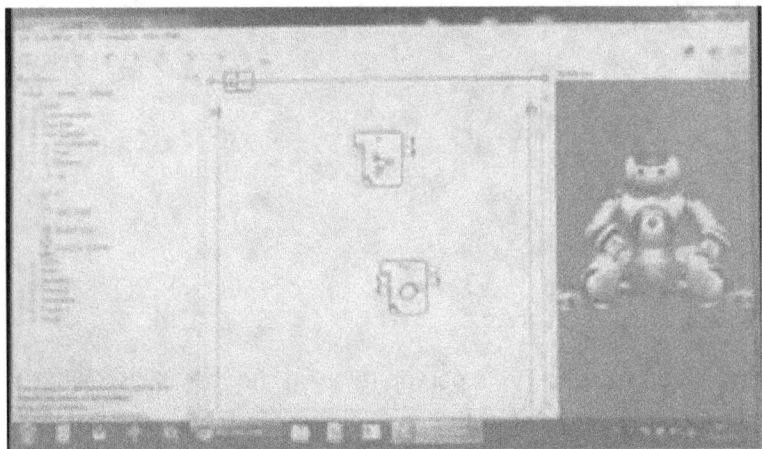

I just shows you the Switch Case, you also have For loop and If. This is for people not familiar with programming, but if you know Python or if you know C++, I'm sure you would better type the for loop in scripts, but here you can create a for loop with some link, like that.

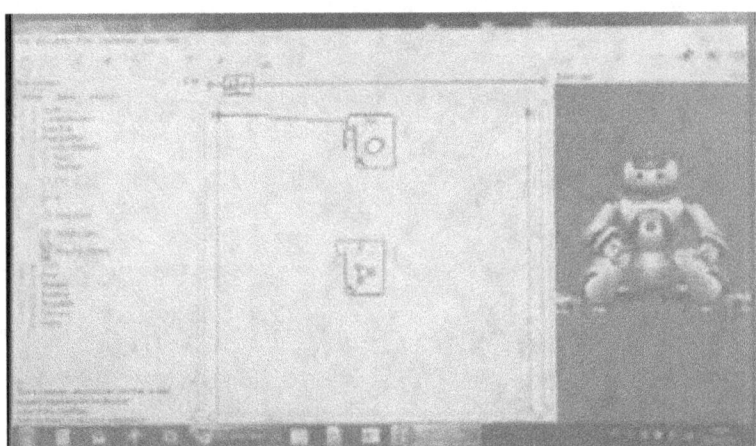

For example, you trigger the For, here it will give you, it will iterate for certain number, so I can link it to the If. How many time to iterate, initial value here, stepping value, final value. For zero to five, step one each time. So it would be zero, one, two, three, four, five and this level would be here. And the If receives the value here, in the input and you can compare it with this condition operator, and you can compare it with value, for example, two. Here you have the Then and here you have the Else. If the condition is True is here and if your condition is False, it's here.

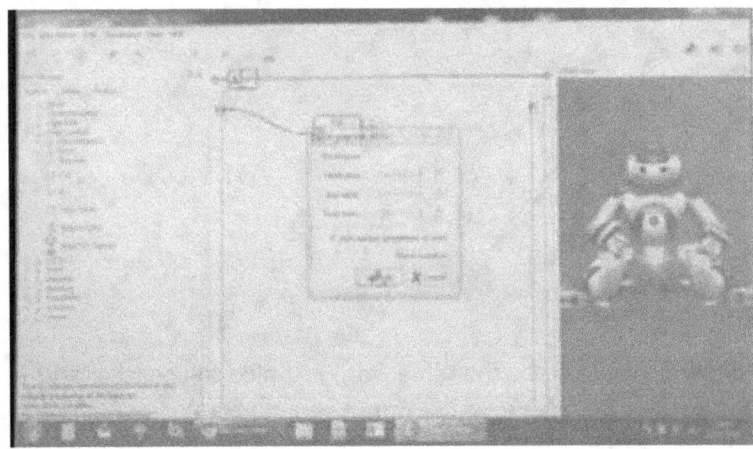

For example, I can make this if it's less or equal than two I can to say "give me more" and if it's greater than two, make me say "enough". And then you can trigger again the next of the for, and you have created your For. "give me more", "give me more", "give me more", "enough", "enough", "enough".

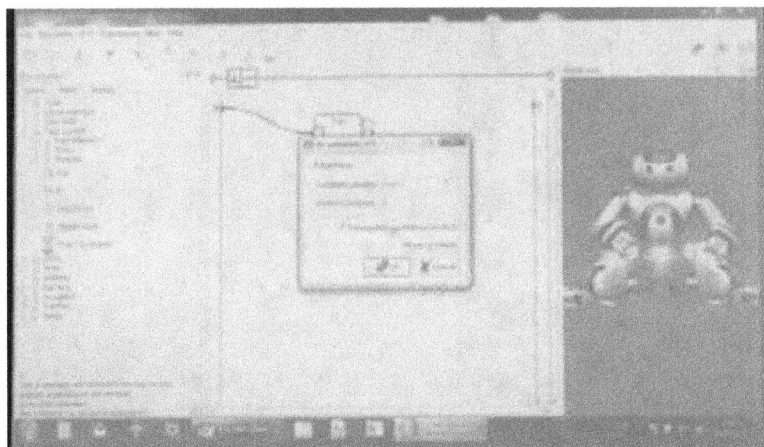

That's great to make visually for loop or If test, but if you are more familiar with programming you can do it with text, with code. For, If, Wait for Signals. Wait for Signals can be useful this box is waiting that the two input is triggered and then it do the output. So it' kind of filter, you blocking for, waiting for two events to occur. For example, if I take Bumper Sensor, I can link left bumper, right bumper and then I link output to Say box.

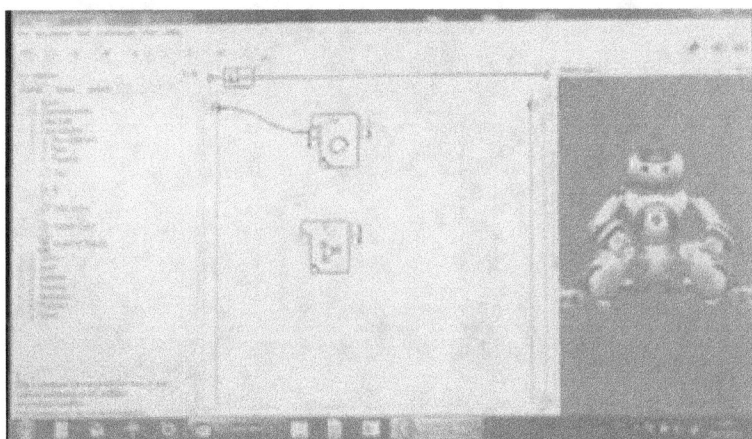

So, when I start the bumper, the bumper box, it will start to watch for the bumper to be pressed. So if the left bumper is pressed this output is triggered, if the right bumper is pressed this output is triggered. And this box will wait for both to be pressed.

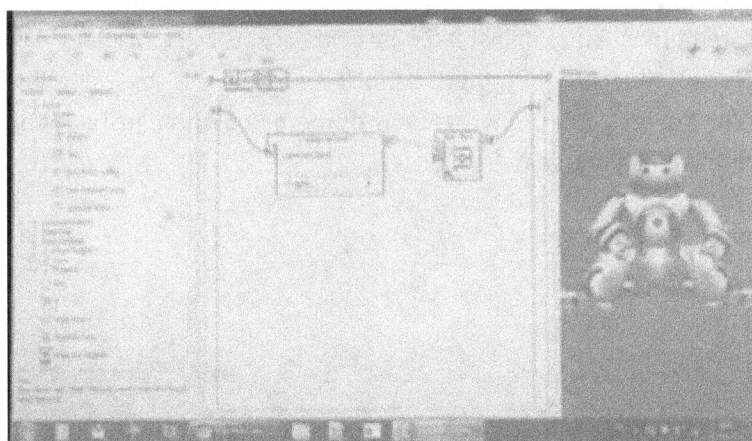

Looks like that. Press this one. I press it several times, but actually it's waiting for the other one. "Hello". And then it's say Hello.

I can redo it, start with this one and this one. "Hello".

 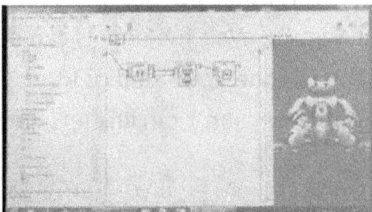 

That's the Wait For Signal. And Only Once is also another kind of filter.

If I just do a direct link like that, of course, if I press several times «Hello" "Hello" "Hello" it will say it here I press three times and it will say it three times. But, if I, instead only Once box, it will block and will just trigger output just one time. So if I press «Hello» and then there is block even if I press several times. This kind of filter, this kind of flow control box, you can write many of them in python, just add to your imagination I'm just giving this one as example, you may find it useful.

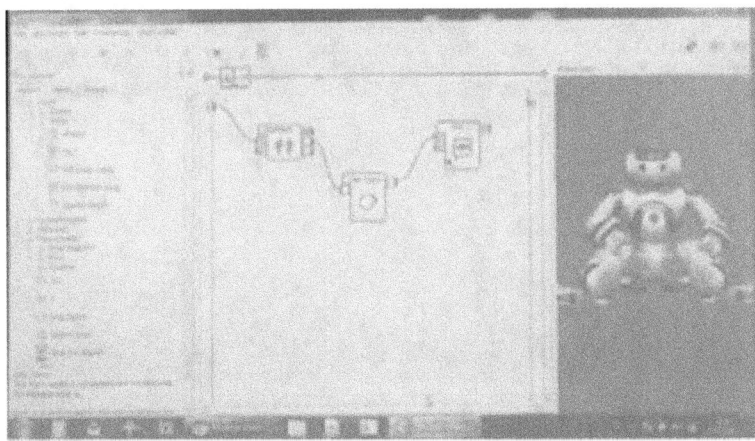

We have a this box I suggest you discuss about it may be different it's to create a state-machine inside Choregraphe. It's this box allow you to control the Timeline.

You start and stop the time inside the timeline. And with that you can control the state-machine. It's a bit long, so I don't want to spending time on this one here. So, maybe get back to this later to discuss it. You also have some very simple box like Timer or Wait. The Wait is very simple, you enter your wait box, you wait for one, two, three seconds - whatever and then the output is triggered, just a delay.

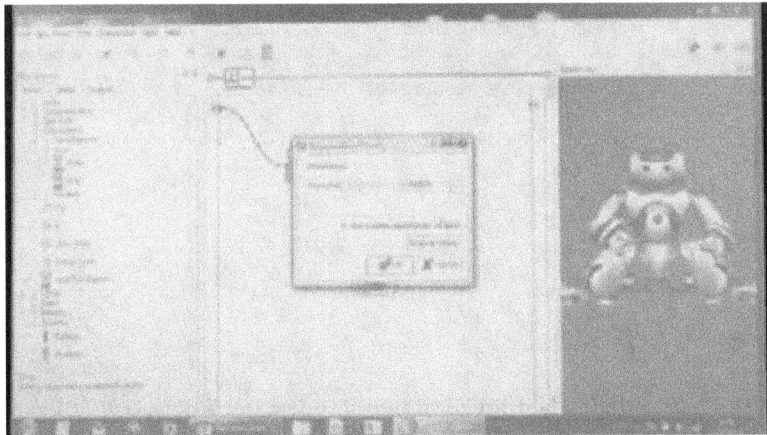

The Timer is almost the same thing; except it will actually send output every setting we having in the number panel. This one, the Get Parameter is to get parameters of your parent box. This demo is quite funny to go, I will do it. For example, I take the Say box. OK, the Say box.

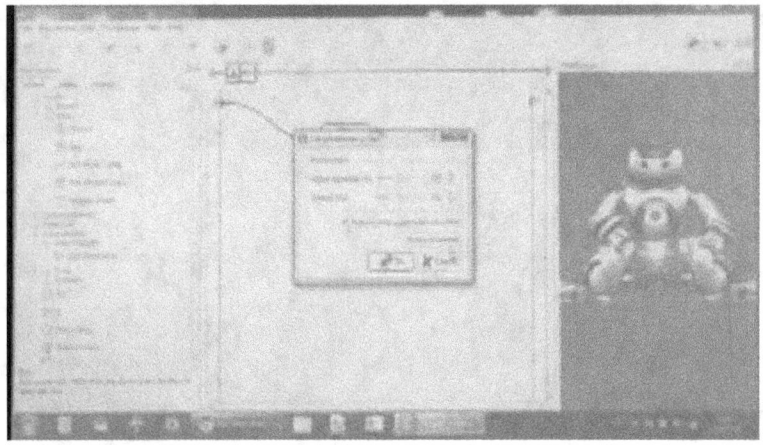

If you look into the parameters we have two parameters for say box, we have Voice shaping and speed.

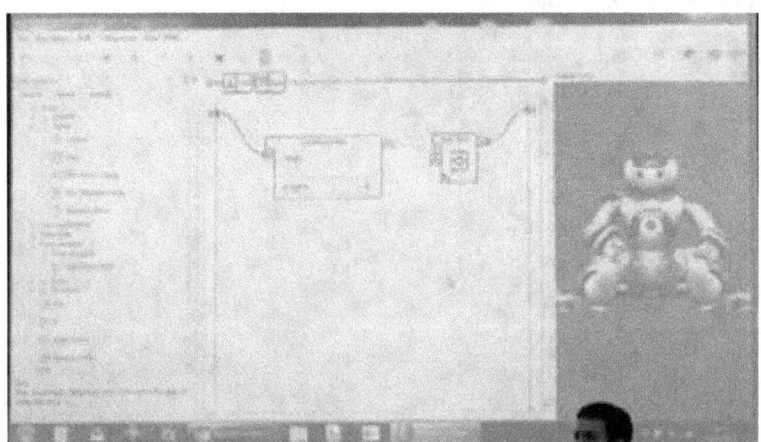

And, as you know, they say box is also the flow diagram box. Here, instead of taking this text, I will get a parameter. Here I will link this. And I link this to the say text. And, actually, I will ask, I will try to get the parameters of my parent box which is the Say box, this one. So, I want to get speed percent.

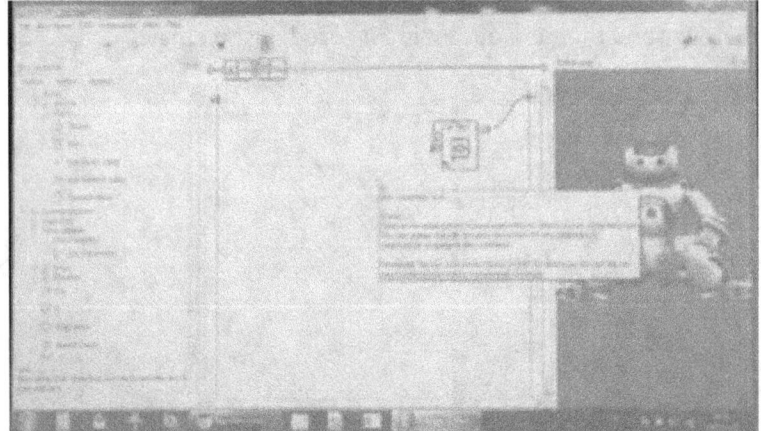

Here, in the parameters of get parameters I type speed (%), that's the name of the parameter for get parameters. I mean, we actually say the value. So if I it say "One hundred" because the value is one hundred, I can make it fifty "Fifty" it say "Fifty" and it say it slowly, because the speed is changed.

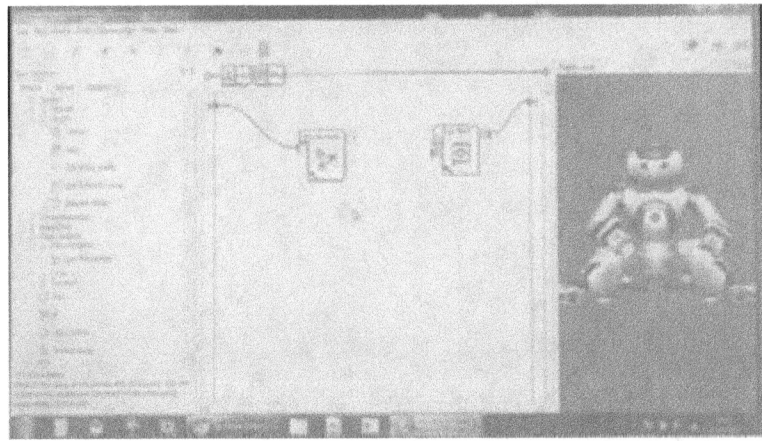

I can make it two hundred. "two hundred", two hundred. That's kind of funny box, we can get the parameters of your parents. Basically, that's it for the flow control box.

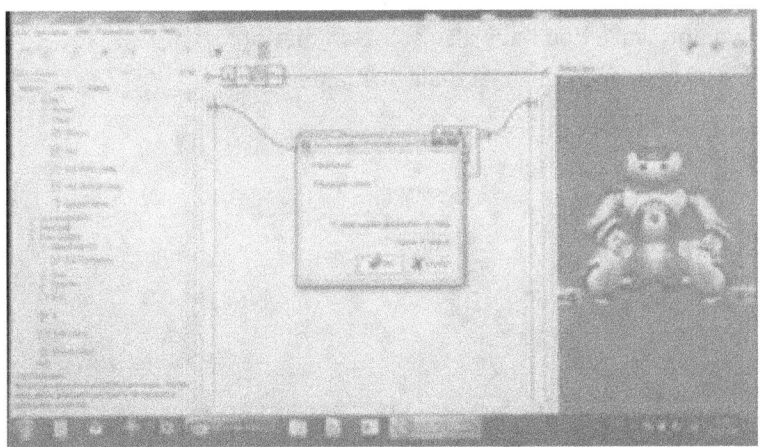

After that we have some box to animate the LEDs.

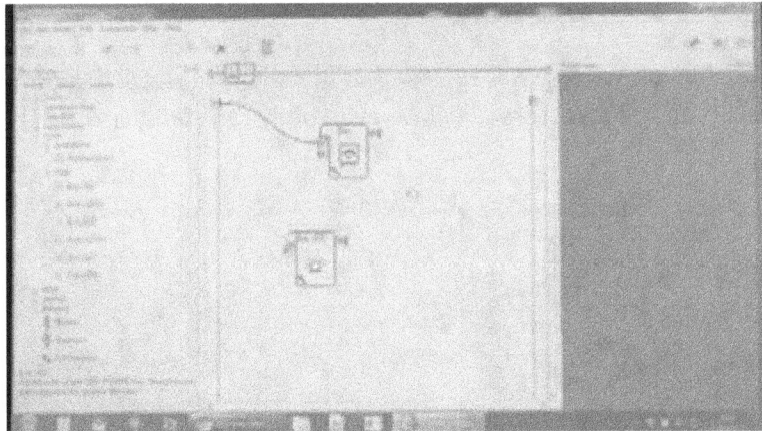

Like, for example, if you want to change the eyes you can double-click inside oops I think I switch off the Wi-Fi, so that' why I lost animation. I kick into plug, the power cord

00:30:00

And now it's blinking I just kick off the stop, the switch off but before it was not blinking. Ok, the robot need reconnect. Robot tries to reconnect. "Hello, I'm robot, my internet address is 192.168.1.103, my battery is fully charged" it changed its IP. So, yeah, basically, the LED box is just to change the color of the LEDs. After that we have some Math box, but it's not big here.   After that, more interesting, is the motions box? Oh, yes, you didn't see the robot moving, yet.

You have useful box like Stand Up, Sit Down and walk. You have also some animations, that I already used, some animation Hello, we have some Dance, one dance, actually just a demo.

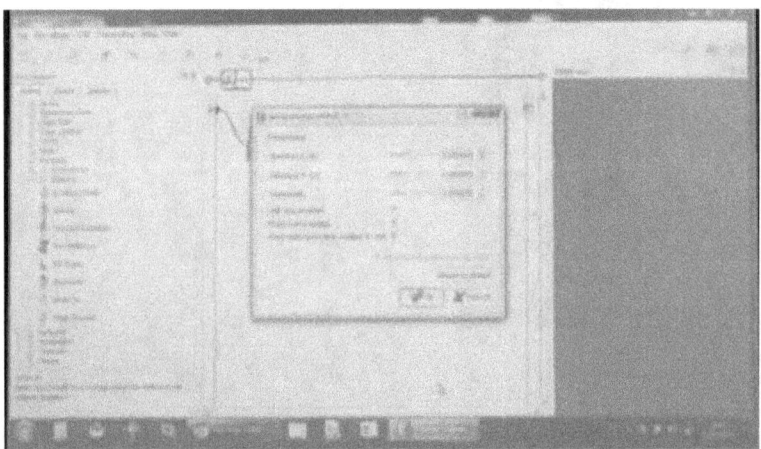

But more interesting is Walk and Stand Up. Maybe I already shown you that Stand up so, you just drag and drop box, it's quite easy, you have one output if it success or fail. So everybody can stand up and then we can walk.

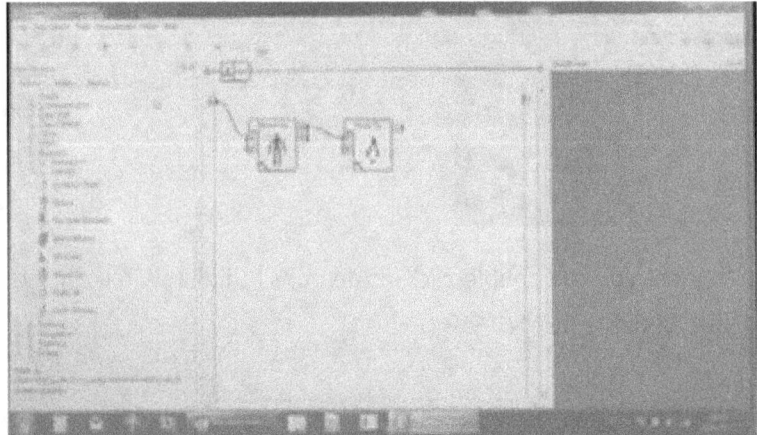

Move To. So, in the Move To parameter, there is different parameters, the X Distance, which is X is in front of the robot, it's like inertial The Y in side and Theta is angle at the destination. How many angle you want to turn at the destination. By default it's just twenty centimeters moving. Let's walk a bit longer sixty centimeters. Waiting for it to boot. [Quiet question from listener] it can stand from many positions? Almost, yes. From the sit down, from the belly, lying down on belly, on back maybe on the side it may not find it, but if the stand is failed, then you have this output which is triggered. So, you can make the robot say "I'm sorry I don't know how to stand up, please help me stand up" or something like that. [Quiet question from listener] Yeah, maybe. Wait. You can wait for a few seconds and then start to stand up again.

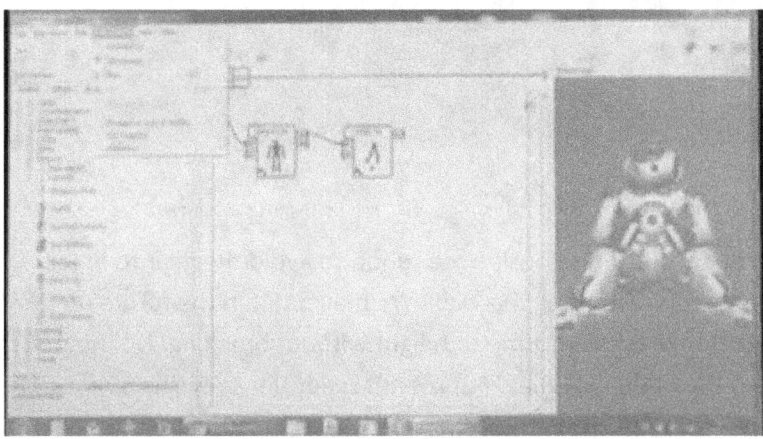

3.6. The simulator robot

Not sure it gave me same IP. [Talking in French] Oh, I should change the language also. Or maybe French you understand? [Talking in French] "Quit" "Quit" " punch my to change " I'm sorry about line I'm changing this Ok, so I'm, I can try my program. When you hold on breaking something, that's the problem. "Ouch" breaks it I can probably not demonstrate it - break this foot. So, I'm sorry, I can't cannot make it walk.

You just show the walk him today [?] You will only see the walk in the simulator, I'm sorry. When he fell down he broke his knee, the knee is blocked. That's why, again, like I told you, when you want animation, you should do it on the floor, or very carefully when he fall down [quiet question from listener] how many broken? The broken gear is quite often problem for us. We have planned to improve it.

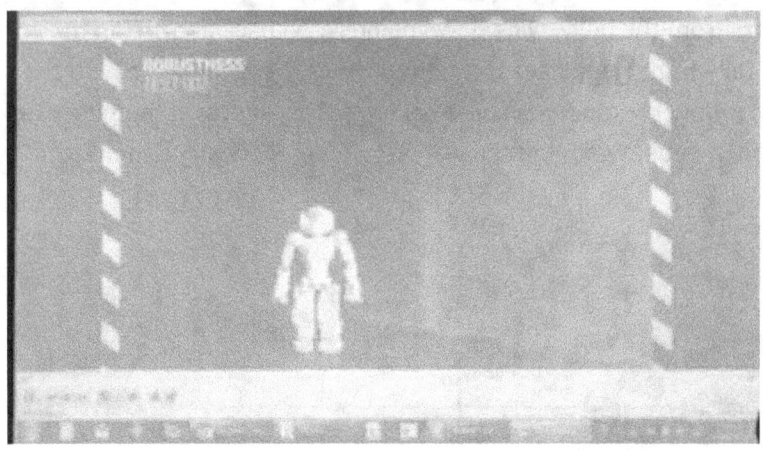

We having plan to improve the to put some torque limiter inside the robot, to avoid the gear to break. But we didn't implement it yet, because this was disturbing the walk. We may put it in the future on robot. And in the future the robot will get to fall on from 1 meter height without breaking. Not now. I will show you video just to just to show you, that I'm not lying. But it's not ready yet so we have removed it, so now it's just breaks his gears for now. [Quiet question from listener] The foot back it's a foot back that move out so, actually it's the same motor load because

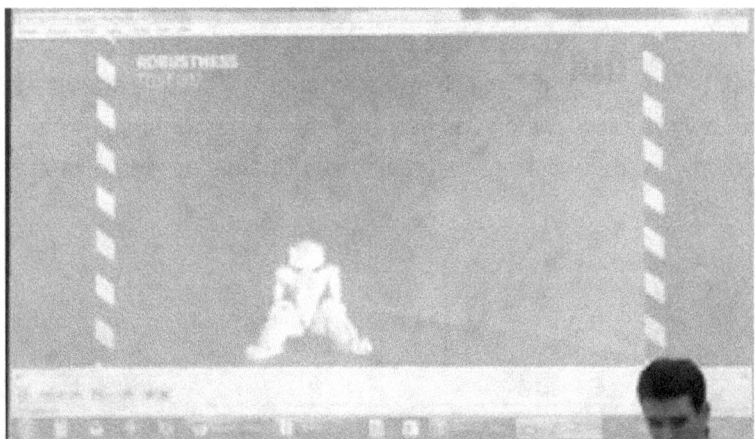

He is trained to move putting around current trying to move foots in the stand position. But cannot because gear is blocked. This makes it more we have. So, don't do that with your own robot. Mr. Innocent robot, or if you want to break the innocent robot. It's not inside the robot, it will be inside the robot, if and so, Yes, you have a walk Move Toward, Move To, but I cannot show you that it in the simulator. And actually it's a good time to show you the simulator, why not, let's do that.

So, the simulator after installing you just launching it, like that And then you can it start like that and then you can press the play button.

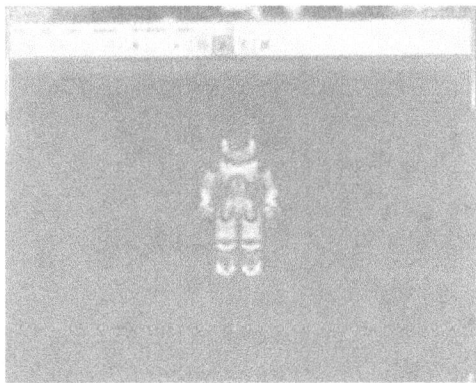

The play button starts the physics simulation and also launches local NAOqi. So you should then you should connect to the local host, so do not use this one, because this one launch another local NAOqi, not the one we running if inside the simulator. So, you just choose local, connect to local NAOqi, if you don't have the simulator, if you want to work inside Choregraphe - it's a quick way to do that.

But if you work with the simulator, you actually press the play of the simulator, and then you use the connect button to connect to the local host.

If you type local host maybe it work with launch. It says I don't have the same NAO connection, because I'm in the window 1.10 NAOsim and 1.12 for Choregraphe.

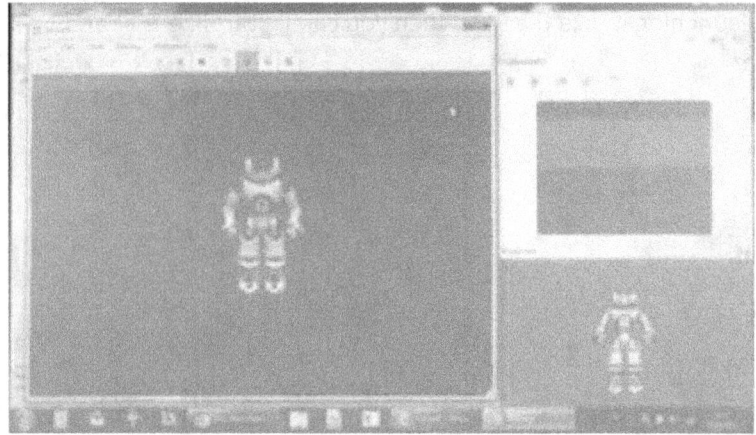

I can open Video Monitor, and then I can test my program. To walk. I guess, that's a different operation. Let's try this one.

Yes, the robot is actually walking, the one I used is endless walk, is just a demo walk, there the robot will just run in circle. I can disturb a little bit the robots [?].

Here you have some tools to pan view, or to move some objects. Or to rotate some object, if you select this for rotating and then if you click on robot, you can actually then if you grab the angle, you can actually like he fall.

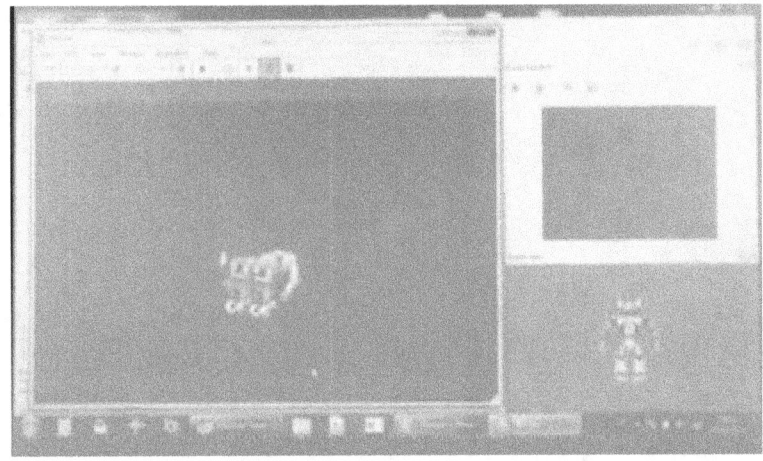

And he will try to move again. And here, I don't know if you have seen this, you have the feedback of the camera from the simulator.

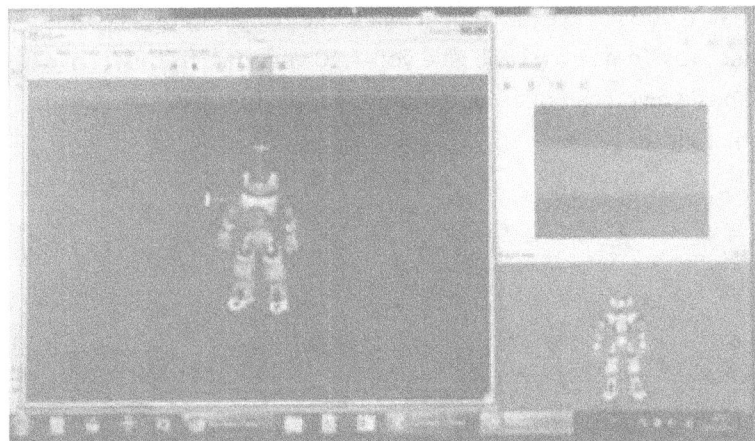

You have the Stand Up box, you have the Sit Down box, which is almost the same,

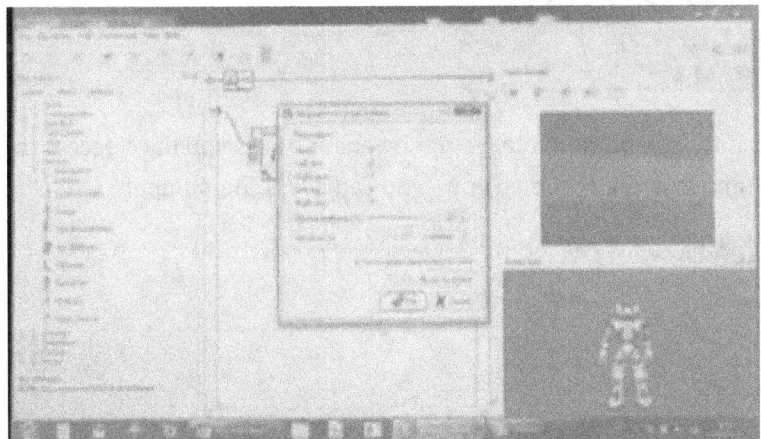

You have the set stiffness box, where you adjust the ratio of the stiffness, or you can get to some fact [?] of stiffness for the robot. And there is Walk To and Walk Towards. I wanted to demonstrate Walk To, but seems that walk to 1.12 animation doesn't work with simulator 1.10. I will use Walk Toward, trying. Maybe tomorrow I can also give you a Choregraphe 1.10 if you want, so you could install it than. The Walk Toward is, basically, quite similar to Walk To, but Walk To - you give a target, and Walk Toward - you give it a direction.

Here you also have X in front of the robot, Y - on the side and theta angle. But for X and Y the value is different: -1 to 1. So 1 means, it's a ratio, maximum speed in flows. So, -1 is maximum speed walking backwards. And same for the Y, which is on the sides, maximum speed on the side, if you put the small value it will make some small steps, like that, if you put bigger value it will make bigger steps, like that. And theta is just theta - angle.   Same boxes idea Tomorrow I will give you 1.10 versions, fix this problem. You also have the box for the hands, so you can open or close the hands. In the parameters you can choose both, or left or right, open or close.

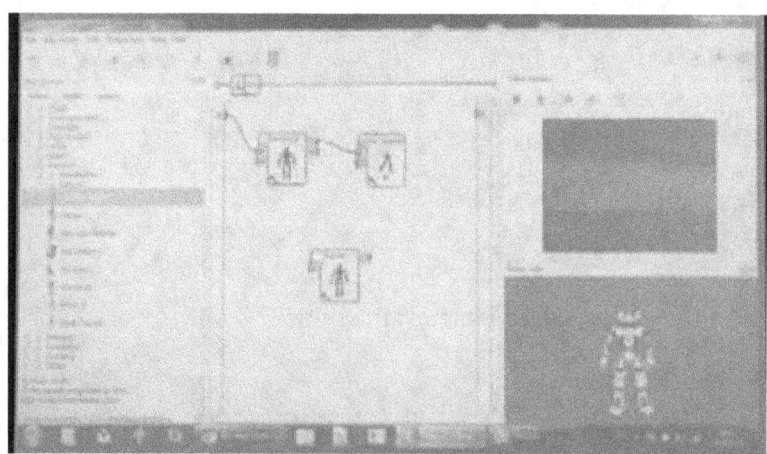

This just a demo, like I shown you. That's it for the motion, really. In the Sensor boxes you have access to many of the sensors of the robot. Foot Bumper, already shown you - you can press the bumper.

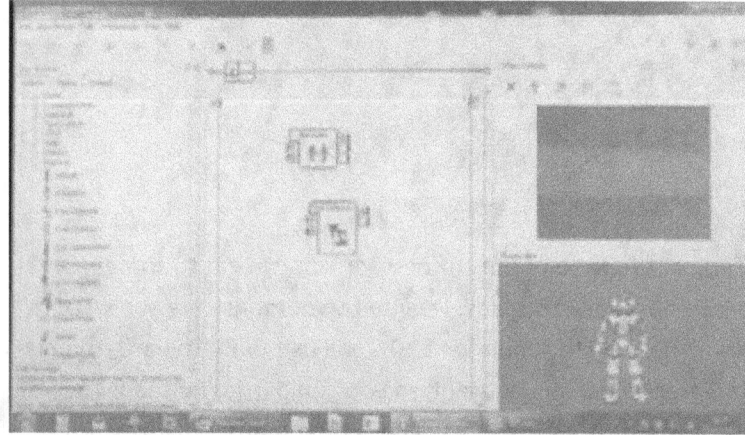

Fall Detector - this one tell you, if the robot has fallen. The Foot Contacts, this one is for FSM [?] here, it's mainly telling you, if you lose the contact or have the contact. It's useful to know if you grab the robot. Get Temperature is to get temperature of one motor, or temperature of the head. Above all the battery level also, so if the battery is low, you have this output, which is triggered. So, you can have the robot say "oh, I'm tired" and then sit down. If it didn't charge, the robot would just switch off, because low battery. Has Hardware input   because we have different version of robot, we have some robots without legs, some robots with laser heads.

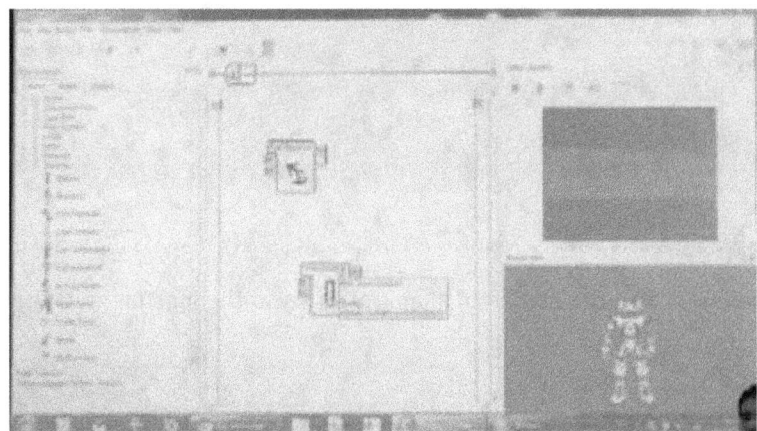

So, you can test, what kind of robot it is.

For you, it's useless. Is In Collision, this one is to tell, if some arm collide with the body, so you can test it, it's not very useful also. The Robot Pose. The Robot Pose is kind of sensor.

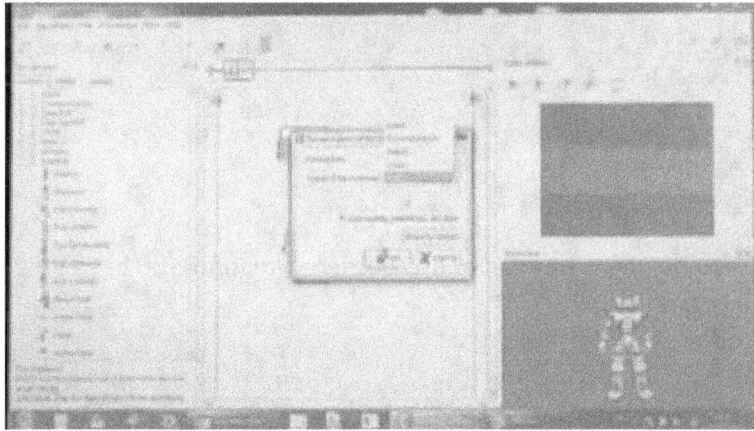

The Robot Pose can give you one of these answers: can be Stand, Sit, Crouch, Knee, Front, Back, Belly, Left and Right or Head Back.

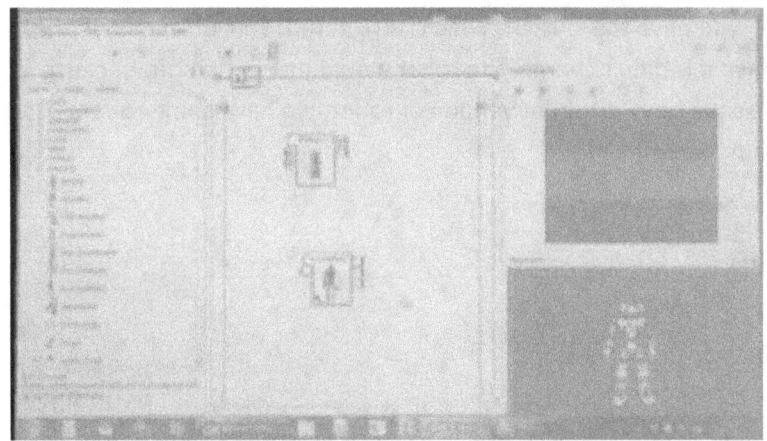

So, basically, robot tries to look at the positions of all the angles of all joints and also the inertial bolt[?].

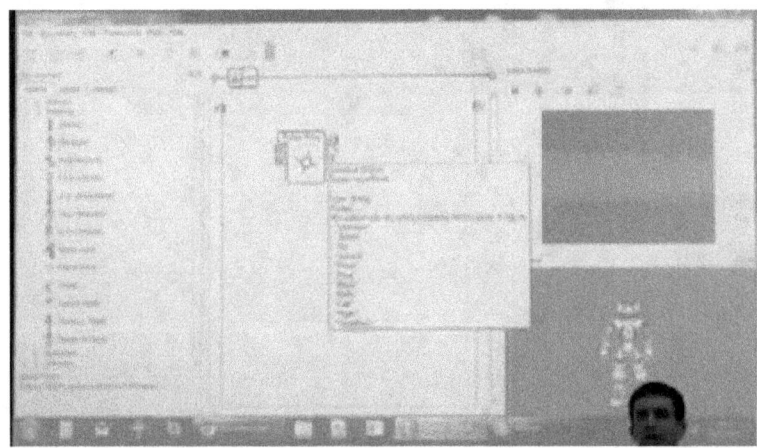

So, it can tell you "probably, I'm lying on the back" or "probably I'm sitting" or on the Knee, like that.

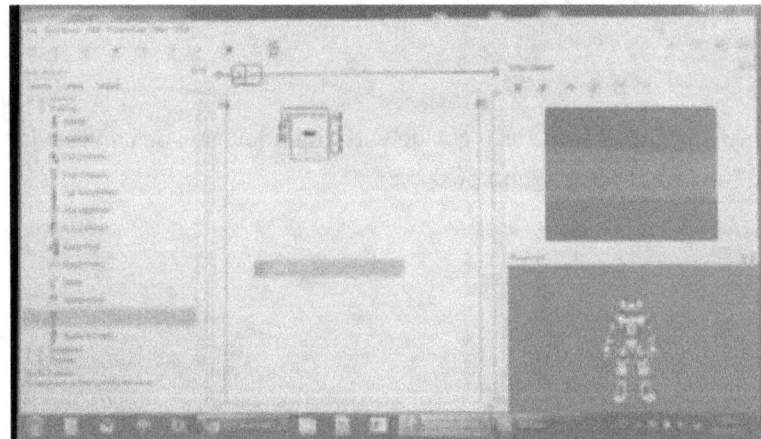

Depending on the position, you can decide to do one animation or do another animation. You have the Sonar. The Sonar is not simulated in the simulator.

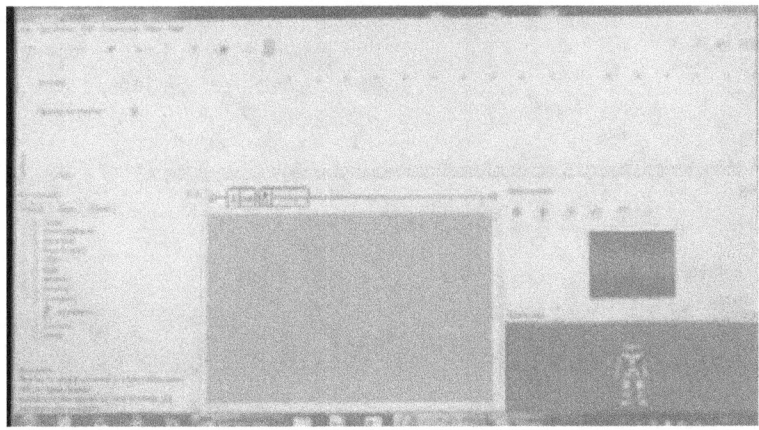

If you have the behavior where the robot tries to avoid wall or obstacle - do not work in the simulator. You have boxes for the Tactile Heads - the front, the middle, the back on the tactile head and same for the hands. That's all boxes for the different sensors on the robot.

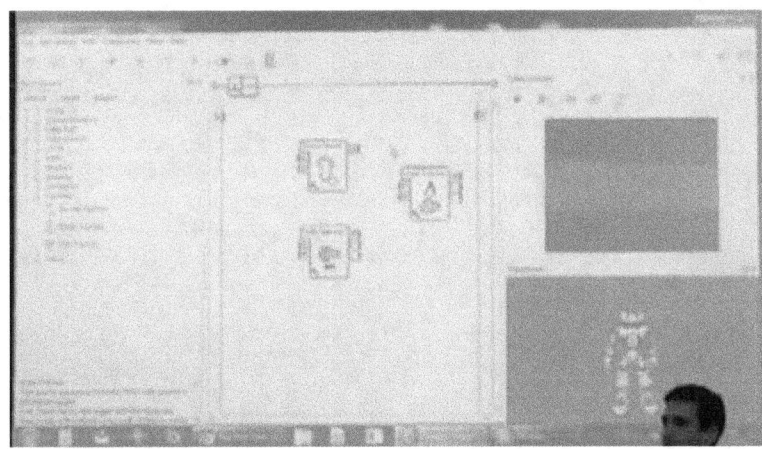

In Templates you just have animation box, this one is just an empty animation box. You can it's to create quickly animation. The Trackers would have I've shown you the Sound Tracker, I've shown you the WB Tracker. The Walk Tracker, my robot cannot walk anymore, but, basically, you show the wall, and you just the wall, and robot toward the wall, keeping certain distance for the wall. And finally, we finish with the vision. This afternoon, I have shown you the vision recognition.

But there is also other box like Face detection. This one, it just tells you if it detects the face, don't need to learn the face. Just tell you if there are one, two, three faces so it gives you a number here. It gives you how many faces it has detected.

But there are also boxes to recognize the face, to learn the face and then recognize the face. And you can also here call the face you have learned.

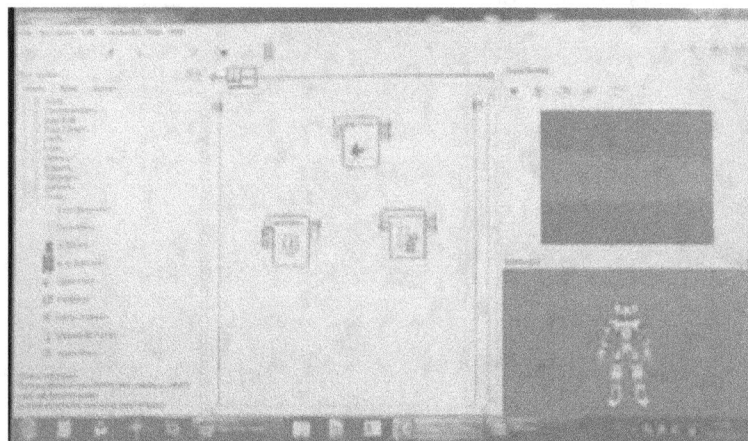

You can try that; also this will not work with the simulator, of course. The way to working is, you give a name. I give my name, for example. And who wants to what is your name? - Hubert. Hubert, OK.

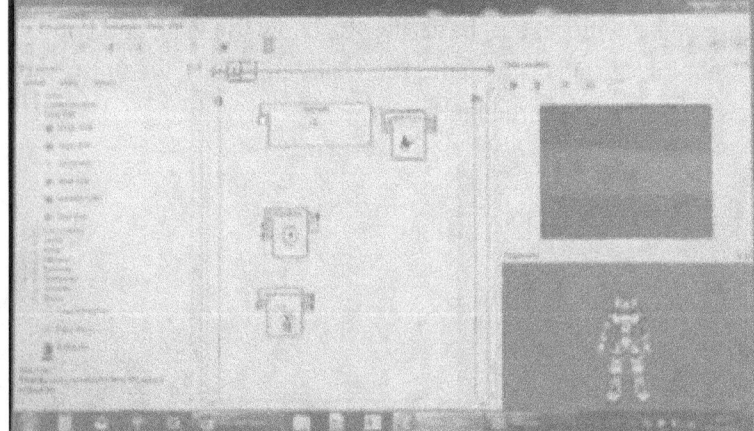

Let's call it Hubert.   English accent impressive. And then, I will ask NAO to say the name as the name of face it has recognized.

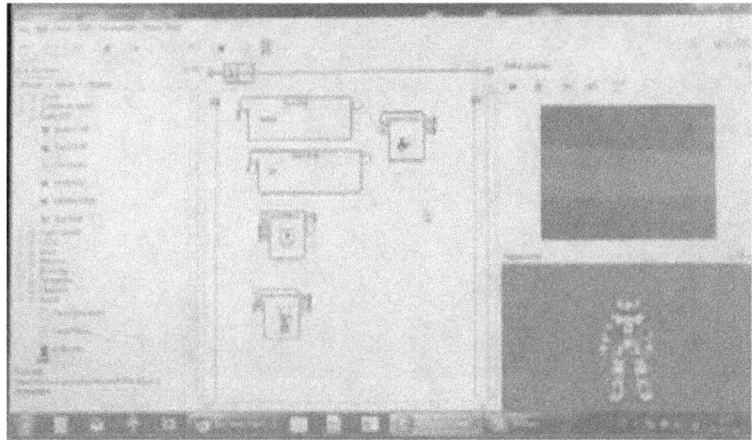

I copied the say text box here. And I will link it, like that. It's a kind of two-step demo. Forget to connect to the will exit to connect to the robot.

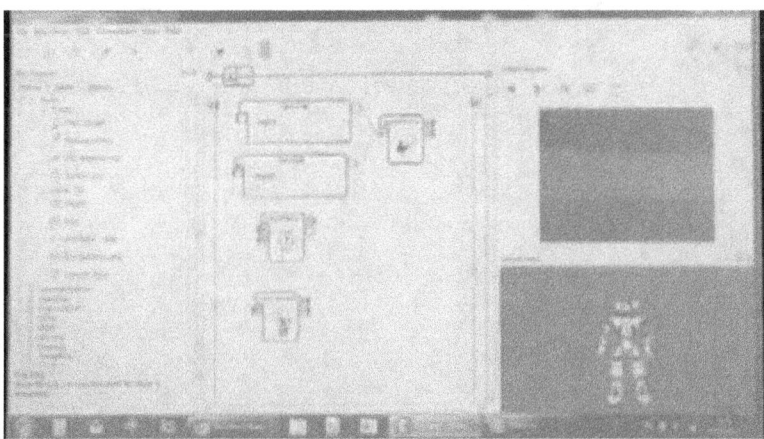

I should avoid very strong lights.

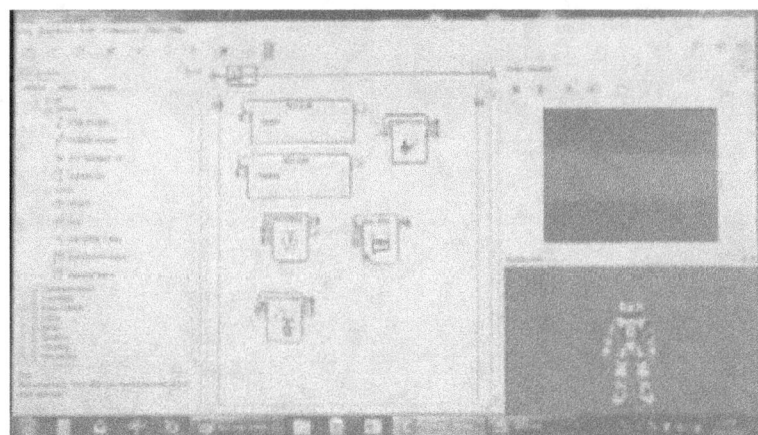

So, I launched the behavior, I did only input to this.

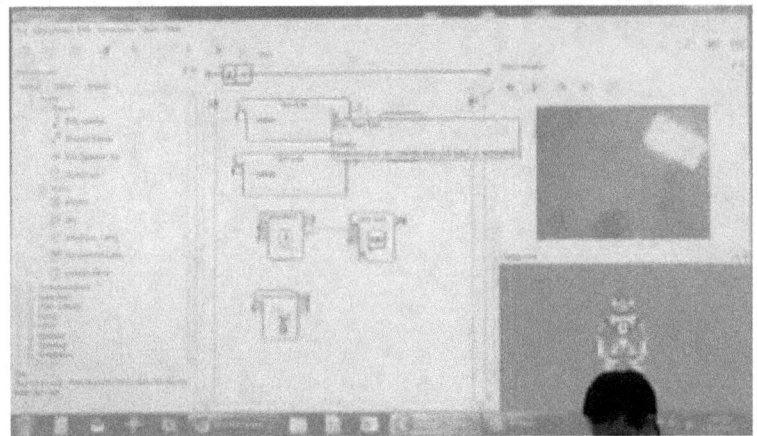

Because, actually, I can just, like, double-clicking any input, I can trigger it. So, I will first double-click my name, it will trigger this one, and then it's waiting for the face. Oh, no wait I forgot something.

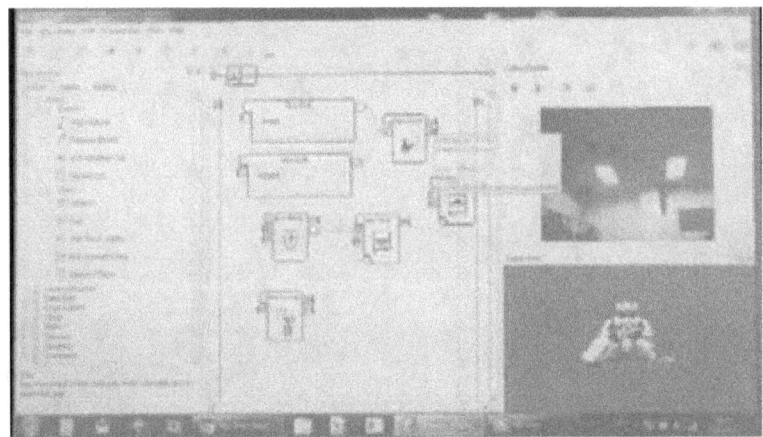

Quite important. You can put to say hello. You can say hello, when you has learn the face.

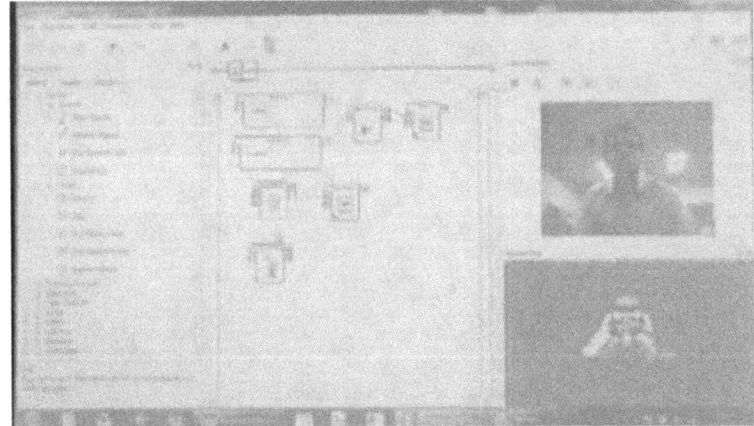

So we know when Recognize finished learning the face. Start again, I'm triggering my name. Will look at the robot. "Hello". And now Hubert, please come I will trigger Huber, just look at the robot. "Hello".

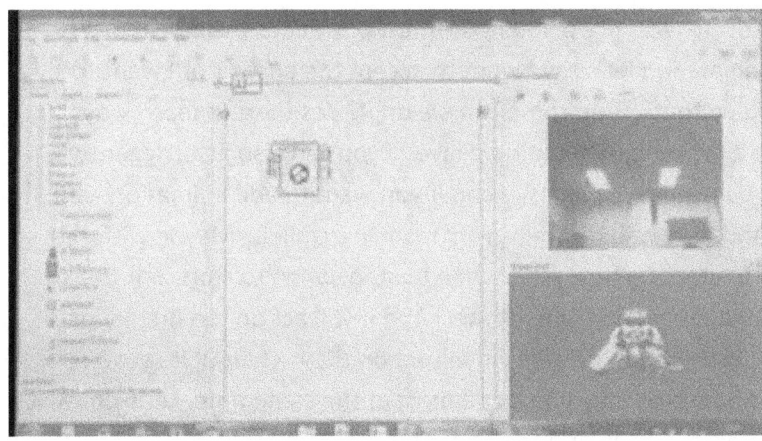

Thank you, and now stay with me, please. Now if I trigger the other box, with face recognition. Ok, you can look in. "Hubert" - that's great recognition. And me «Albert" - he doesn't like its master face. So, basically, it's working like that. The face recognition is based on characteristic points on your face. For example, the distance between your two eyes, or the nose and the eye and the mouth.

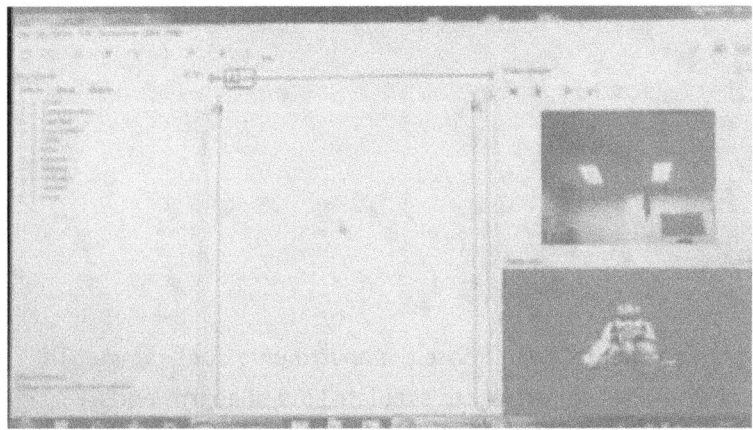

All this kind of information on your face. The robot doesn't store the picture of you; just store this biometrics of your face. And then, when it detects the face, it measure for this biometrics stuff and try to search in his database which one is the closest one. And when it finds the face, which is closest to the current one it just tells you, that it has recognized this one. Basically, that's it for the just about this one basically, that's all for the vision boxes.

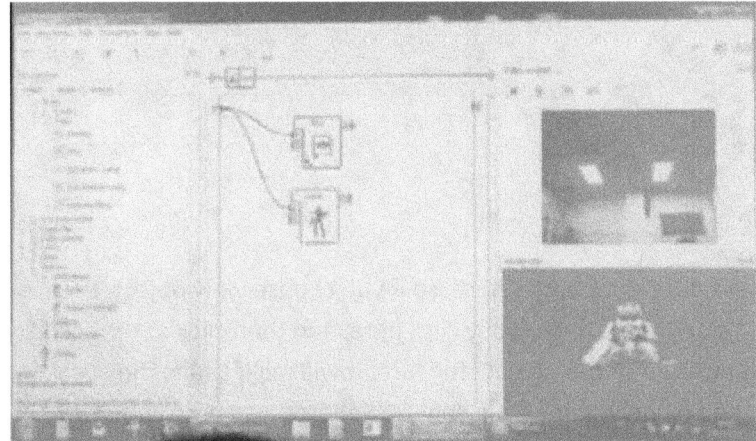

Oh, yes, there is the NAOMark. We need NAOMark, just trigger it, and it just give the number here, if you have recognized it. You can use the NAOMark box to easily make some paper [?] to show a mark and when you have a number of it recognized. This one is to select top camera or bottom camera, depend on the trigger basically, that's it for the vision. Do you have questions about the box library? [Quiet question from listener] timeline animation let's say I want to accept the answer or I want to touch my head at the same time. Can you do that? Timeframe can we do it? Yes there is many ways to make parallel behaviors run. You can use a timeline with multiple layers; you can also just trigger two boxes at the same time, for example. If you want to have a say and if you want to add animation, you can just trigger them at the same time. And they will, basically, run, mainly, parallel. Maybe with the timeline you have better control on how to stop, on how start when to stop, when to start. But there is many ways of having parallel control. [Quiet question from listener] And it will act on the first input it sees? I cannot guarantee sorry? [Quiet question from listener] it will act on the first input it sees? Yes, I think I cannot guarantee which one will stopped exactly first but almost at the same time. Let's finish by more stuff on NAOsim. There is Play, Pause simulation and the Stop of the simulation. You can also create a new a new environment.

We have empty environment, basic room or apartments. We choose the apartments, for example. The apartment is quite simple apartments, but if you want to test, try to simulate kind of apartment's interaction with the user, why not - you can try to use the apartments.

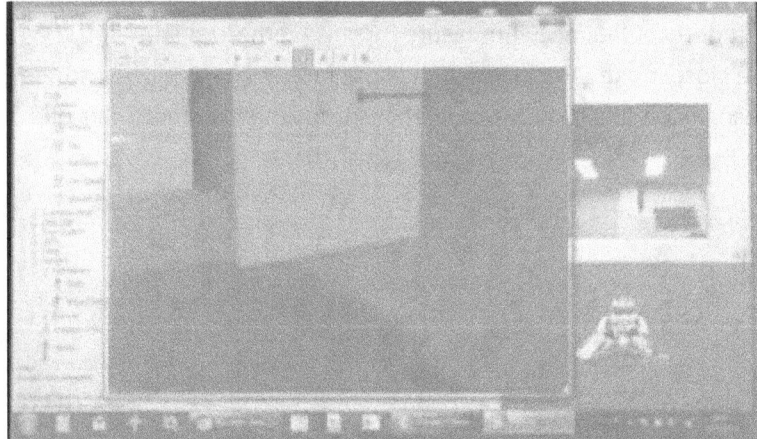

You can also have some objects in it, there is different objects here, so if you choose view object, then you can drag and drop some object, like this one, you can actually click here, and then click in the scene. Then you can choose this button to adjust position of the object for sure. It will fall if I start the simulation, the object will, of course, fall. And by here some little sphere. Spheres also drag and drop here. So, you can save this, and I guess you can also create your own scene but I'm not sure about that.

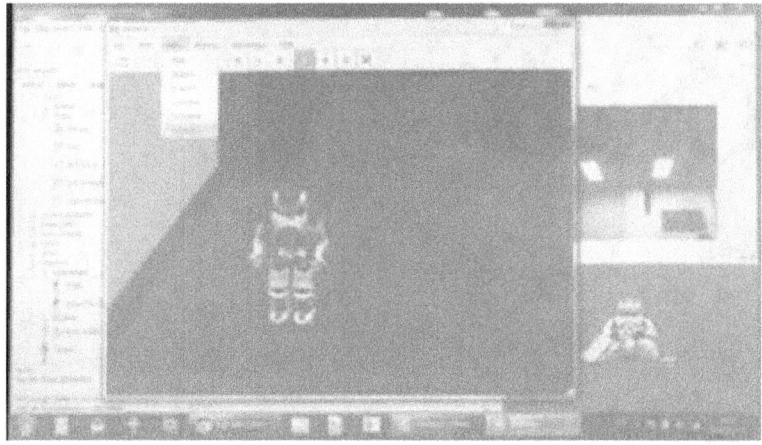

Basically, you can do that. You have explorer, can allow you to select the object in the scene. If you have many objects, you can quickly connect to select it. You can zoom in or zoom out with the mouse.

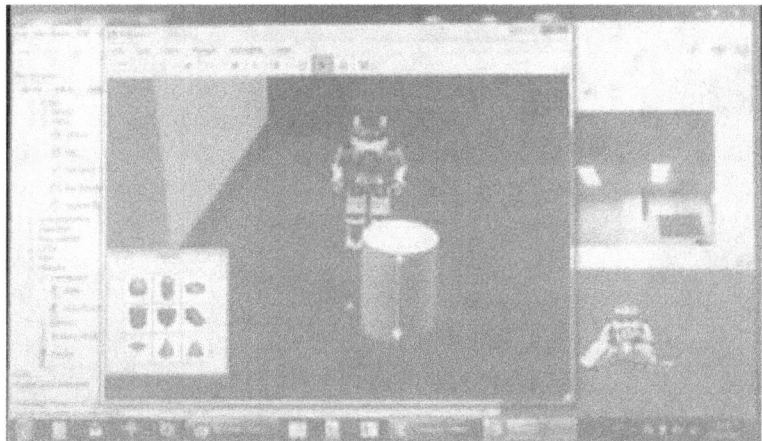

Statistics is not very important, just a frame rate. We have a you can have different camera of NAO, but the camera of NAO is also not very important, because you also have it here, in the Choregraphe.

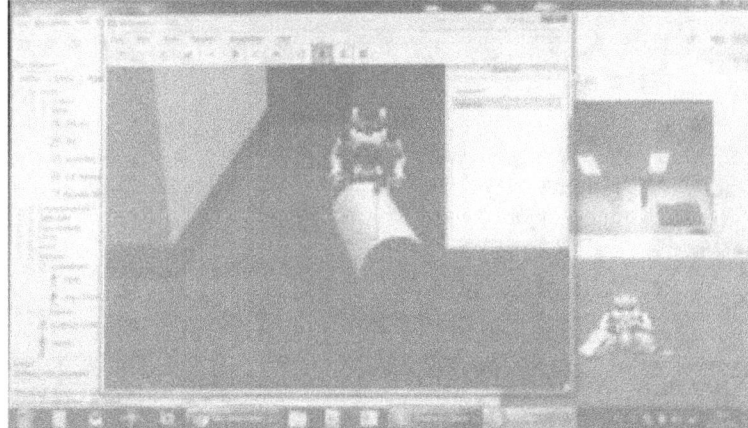

Not very important. And some properties to place the object, so if you select this object and if you want to place it at the very precise position you can enter the position.

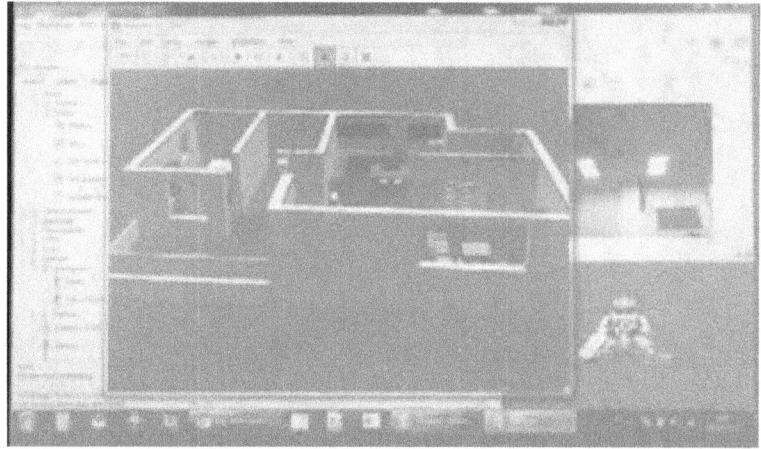

I changed the color of the object, so if you have a some vision algorithm, that uses a different color of the object, you can simulate it here.

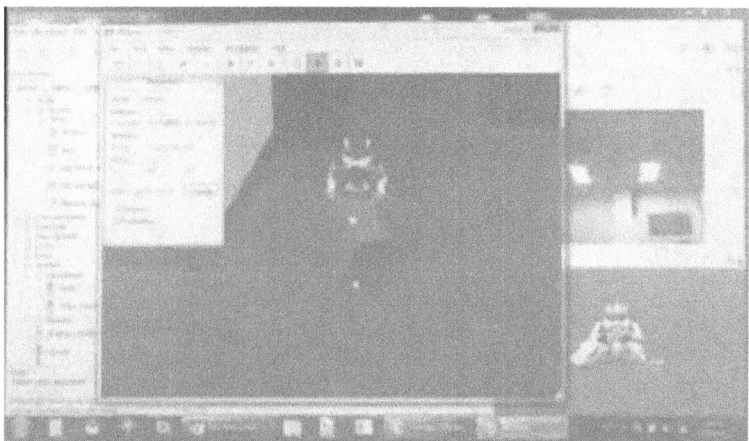

Some options for the physics. Rigid body. Not very important and, basically, simulation stops. Basically, that's all you need, because the simulator is just the simulation of the real world, but mainly you debug tool is in the Choregraphe or in C++. It's just to see how it's may look like. It's not important.

You can also debug your program without simulators connect. If you have a if you just want to debug some logic, you can just drag it, like that, with Choregraphe The simulator is useful, if you have animations, you create animations yourself and you are not sure if it's stable.

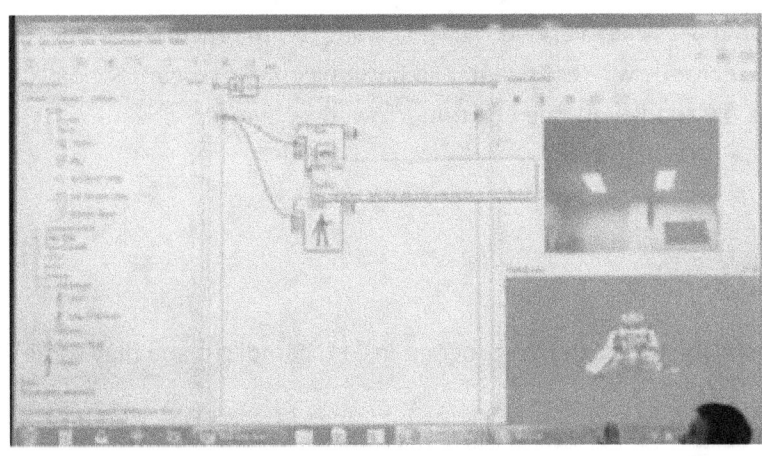

So, all you want to move exact parts important to reorganize the component. OK, what else can I say about the NAOsim, that's before NAOsim, that's why I begin take too many times, because, as you can see NAOsim is quite simple software but that's enough. So, do you have questions ask questions about the day?

01:05:00

[Quiet question from listener] can we take task a box like that, with C++ inside? No. This box have python script, because it's a scripting language, so you can code program, do the robot language program it's interpreted on the robot. For C++ I will show you tomorrow morning, so tomorrow morning we do the C++ programming. And for C++ programming we need to compile it. We need to really compile it to binary... to binary executable or library and then you have to move this binary library or executable inside the head of ... or you make a... in C++ ... but you have to compile it. It's not as easy as scripting programming. Python scripting is a script, there is a virtual machine, there is python running on the robot ... python, python interpreter integrated on robot and it will parse your code... execute compiled code on the fly. [Quiet question from listener] Yeah, tomorrow will show you - you can run your C++ program on the robot and have a connection between the robot and the PC. That's easy to debug and yes, it's very easy to connect like that and you can debug on your PC. And when you finish debugging if you can compile it... and move it to the robot. And the API is the same... it's only a matter of rebuilding ... [quiet question from listener] Yesterday, you mentioned something about your server... Yes. [quiet question from listener] What is ... So, you will not need it. You don't have to use it... Basically, what we have is ... is, we have SMTP server and this is to ... to the device connect to robot... here robot and device have not the same application ... here you can connect with Wi-Fi is ... smartphone ... But you can have your robot here and someone ... on the tablet ...very far and you can connect with SMTP server on a robot and can control robot ... That's what I want to demonstrate, that's why we have SMTP server. For you - you don't need it. ... So, do you have other questions? [Quiet question from listener]... Fourteen centimeters per second. Fourteen, yes. Now, I will refer it to if I find it for driver... I will steal it from it. Fourteen centimeters per second, it's not very fast, it's not very fast [quiet question from listener] ... Basically, I need to get one gear to get it to robot [quiet question from listener] you can install your gear to robot? Yeah, definitely Usually, I take my toolbox with me and some gears, but today I forgot to take it. I just but there is many robots and you No questions? No? So, tomorrow, C++ programming, I will explain NAOqi and introduce C++ programming. Then, in the afternoon, we have we can have open questions and some exercise, practice. So, you can already think try to think about which project you want to do. Because then, at the afternoon we can start to implement it and you can discuss it, actually. So, please prepare some questions. And the group is not set yet. If you can have, if you can found some group already you can start working on your project. I don't want to make a too long   already on this already very heavy. I just want you to practice and do some sort of actions. Everybody will come tomorrow? Yes? Great. Sunday you come?   I mean, Sunday is again, maybe if you want some questions OK, thank you, see you tomorrow. Bye, bye.

4. Robot programming
4.1.Introduction to the architecture of the robot

Today, this morning, I will introduce you NAOqi, it is better to program in C++, compile it and play some example also with local NAOqi simulated NAOqi.

Basically, like I said yesterday, NAOqi is the event Main Program running on the robot controlling all the motors, all the sensors and while you program in C++ or python, usually you make program, and usually you make library for NAOqi.

So, this morning I will introduce you concept of: broker, module, method, proxy and the difference between local and remote module also concept of binding, binding method and, maybe specific And I will show you how to create module in python or C++.

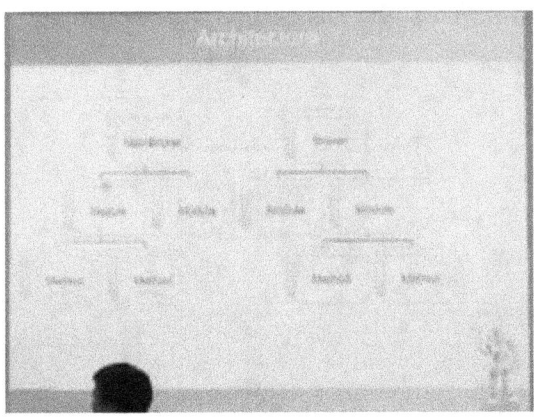

4.2. The basic procedure

Basically, architecture is like that. You have the mainBroker, which is a executable, and this broker contain several modules, actually it loads different modules - libraries, so it loads several modules and each module contain several methods.

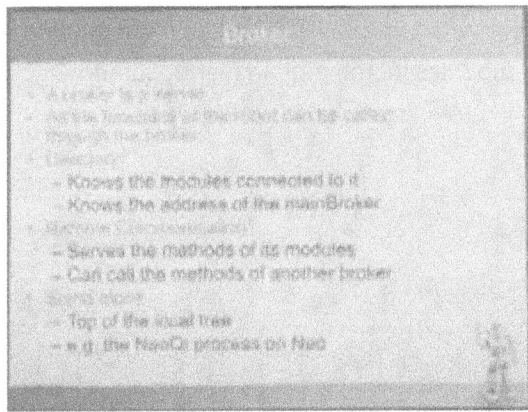

You have mainBroker; you can also have other Broker with exactly the same objects. A broker is, you can say, a server. And all the functions of the robot can be called through the broker.

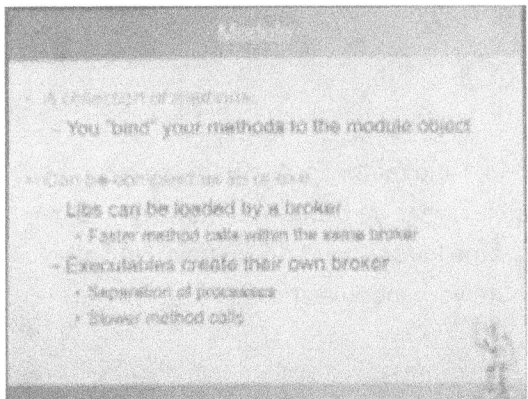

Broker knows all the modules which are connected to it, of course all the modules. And broker also knows the address of the mainBroker, so if it is the mainBroker, the mainBroker knows that address otherwise it knows the address of mainBroker. The broker can do multiplication, it can serve the method of, the method of its module and it can call the method of another broker.

4.3. Local module

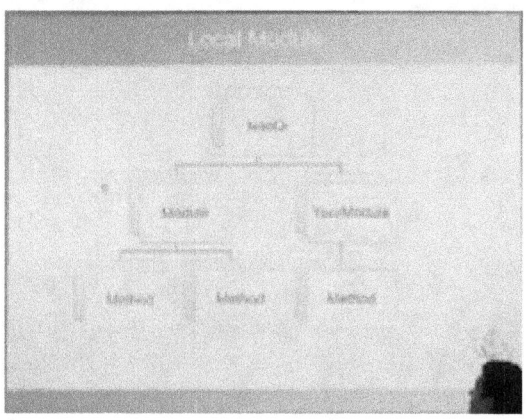

Like I said, the broker is executable, so it standalone program on top of local tree. And, actually, NAOqi, main program, this is the mainBroker, the one process on NAO so, module. Module, what is a module? Module is a collection of methods; you can really see these in modules. And, when you compile your module, when you create your own module and compile it, you can compile it as executable or as library. If you compile it as executable, actually it will create its own module. And if you compile it as a library, then it can be loaded by NAOqi. Of course, it its library, method calls faster than of exe process, but if you compile it as executable you have separation of processes, then it's a bit slower. What is the difference between local module and remote module? Basically, what you call local module is a module that is loaded into NAOqi. So, NAOqi, then broker, as several modules in, the entire layer or something [?] load method, and you have your module, of course with all methods loaded in NAOqi. Is the local module.

4.4. Remote module

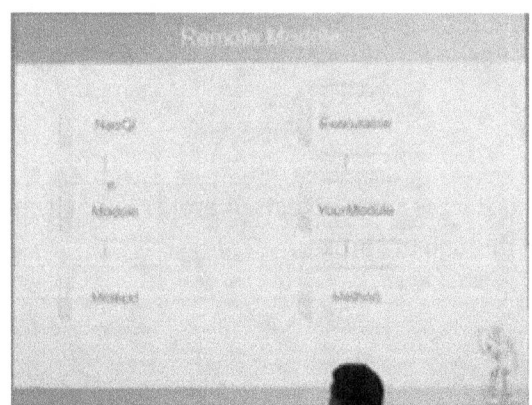

And remote module has separation. Process NAOqi loads module and methods. And you're executable and you're Module and its method. Be careful about the modules what we call a local module is not network serve [?] For us, we can have a remote module on a same Pardon this So, you have NAOqi running and in NAOqi you can also have your code exit the route your module and we call it local module. You run this inside NAO.

4.5. Methods

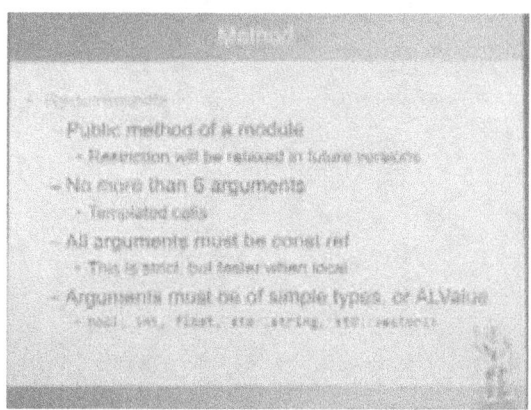

And, of course, you can have another director here you can also have parameter module and, if you connect it here, it also will work. This one is remote module, this one is local. The local module is really the one loaded into NAOqi. Visually that, of course, when you compile your module when you compile your module, if you compile it on your laptop. Which is running Windows, for example? So, it compile windows executable, but if you want to compile it for NAO, you need to have Linux executable Compile your code under Linux it only support so, the Method. Method inside the module. Method must be a public, and you cannot have more than 6 arguments have template calls in modules and argument must be const reference. So, I'm talking about this method, I'm talking about the method you want to expose in your module. [Quiet question from listener] the local module we have method are we have that's a good question. Actually, if you have a simulated NAOqi on your PC, when you can also have local module loaded into your local into NAOqi running on your PC. In that case, NAOqi is windows executable and local module would be here. From here to here. And here is you won't see this from here, but

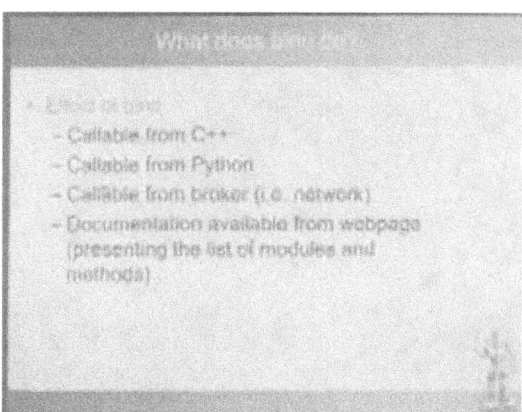

So, yeah Here is an executable, here is an executable, but on Linux you don't have extensions and here is basically, what we call local module is just module compiled library loaded into and it's faster. But remote module also have some potential [?] It's mainly because it can offer you remote module reading on your PC and connecting to a NAO. So you can have flow process in the NAO sensors, and you can debug it on Windows. [Quiet question from listener] What we have to do to we just have to know method? You just need to know IP address of your robot. After that it happens exactly the same I will show you this. [Quiet question from listener] Yeah it's mainly used in a remote useful for debug purpose. It's a bit difficult to debug and trace your program inside.

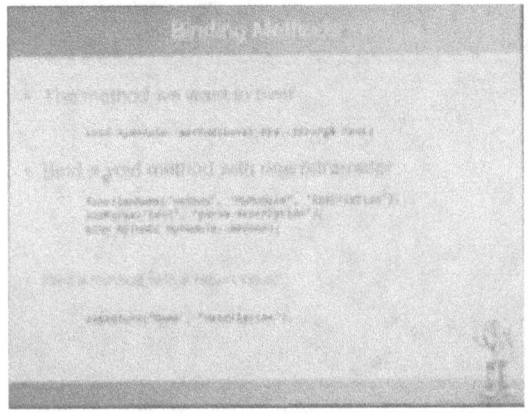

So, yes, I'm talking about method inside your module, must be const reference, arguments must be const reference and arguments must be simple types, ALValue which is , or bool, int, float, string or std::vector. Binding now. What do we call binding? Basically, binding is making your method callable from C++, python. So, basically, you make your method callable. And that means this method can be called also by the broker that also means that documentation is available for your method. A few libraries today, I show you the NAO page we have, all the list of methods access to documentation. So, this is the same for branch module. Oh, thank you. So, yes, how to bind a method. Basically, what we call here is a macro BIND_METHOD. So, imagine you have method, called Method, in your module. OK, so this is the name of your method - Method. And you have a module class, which is called MyModule. Your method has parameters, which is text. And what you do is you should call a function name, to describe the name of your method, the name of the module and a description, which is documentation for this method. Then you can also call addParam function to also add documentation of your parameters. But, like I said, what is important is BIND_METHOD macro

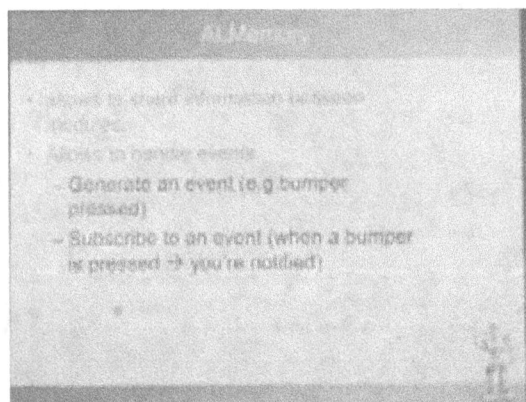

And you actually give a reference of your method. And if your method here, the method is void, doesn't return value, but if your method return value, you can also set return function. So, ALMemory to mention it, there is just a little So, ALMemory is also very useful to communicate between modules.

4.6. Proxy server

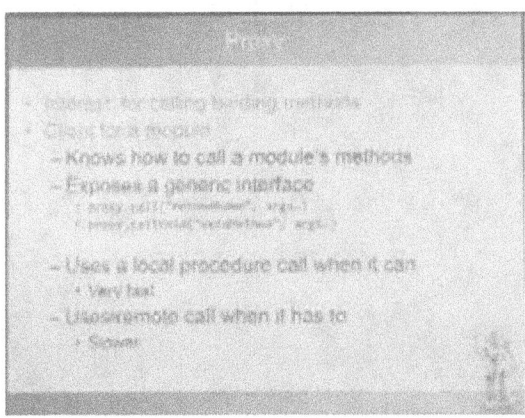

So, you can store some data in one module and another module can read this data. Usually, its preferred way to communicate animation [?] pass animation one to another and you can subscribe to event. [Quiet question from listener] No, you can you can register subscribe to event callback to see it's changed. So, that's why, yeah, that's why The Proxy now, OK. So, we don't directly access pointer of the module. Which Module. Instead, we use a Proxy.

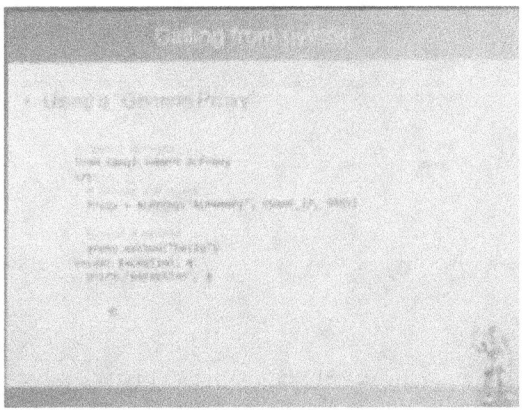

Such as, if only your module is this, not available, then you program may not crash. For example, if you imagine, you have a remote module running on your PC. And you have a program running on your on your robot. Program, I said - your module. But if you suddenly stop the module on your PC - program is terminated [?]. So, that's why we need access to module through proxies. And the proxies know how to call the module's methods. For example, if it's a local module it need a certificate [?] call, if its remote module it's [?] and it should do it through network. Imagine that situation - through Windows, through the network, but on the local host. So, of course, all for local procedures is faster And the proxies exposes a generic interface to call from method. So, the interface is like that - Call, then you give a method name as string and then you give arguments. And that's for a method, that return some [?], and method that doesn't return use callVoid. So, that's why, actually, the method so, how does it work in python. Just use a function ALProxy and then you give the name of your module, for example, here it's ALMemory. So, that's ALProxy on ALMemory. Then you get proxy online [?] which is some indirect [?] pointer to a module, it's encapsulated There is also two optional parameters. The robotic and the port. But if you don't specify these parameters, then it assumes, that you asking module of the mirror [?] of the robot. So, the program is working inside that robot, inside just proxy of module

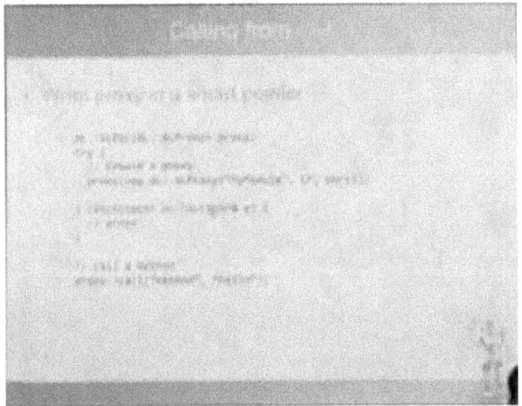

And then, once you have your proxy, you can call oh, sorry you can call your method just by using ".method", that's the method name as parameter. Because its python, so, it's a bit easier, you can directly write it like that. For a C++ it's a very similar, the syntax is a bit different. C++ syntax.

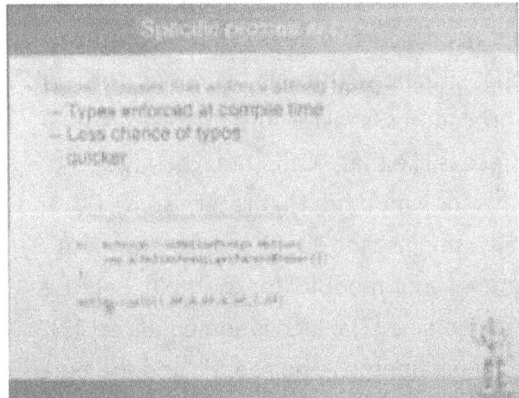

So, this time, you create a pointer on ALProxy. Base class ALProxy it's in AL namespace, and proxy pointer. So, if you remember, what I told you yesterday about python - I told you that variables are not typed, that's why there is no type in this variable. No type for variable in python. But in C++ variables are typed, so, proxy, you create a new of the class ALProxy, you give the name of the module "MyModule" then ip, port And then just call, function call, give method name and a parameters. It's very similar.

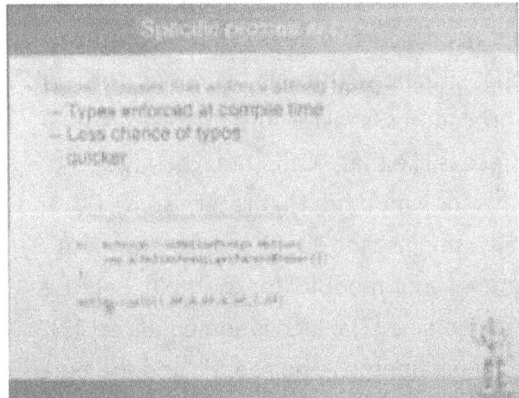

But now we can also convert [?] a bit more, in C++. To be more convenient. This is what we call Generic Proxy. Because you can have this proxy, it works with any kind of module, even your module, if you write you own module and defined your method, you can use this proxy. It's generic, of course but we also provide some specific proxy. To better [?] explain specific proxy, I will just make a little remark about smart pointer. Smart pointer, anybody knows what is smart pointer? In C++. Yes, you know that? So, maybe C++ experts know what smart pointer is. Basically, the smart pointer is a class. It's complete

[?] class, that contain [?] simulated pointer. It's in the class reference which means, every time you want your smart pointer on one object, it's just direct[?] this and when the reference count goes to zero for that object, that's why it's called smart. So, basically, AL it's a class and it's provide a smart pointer that you can find it anyway in application. Smart pointer. We call it ALPtr. ALPtr - the smart pointer on ALProxy. So, that's instead of a real pointer, which is just pointer. And after that it's called, we say create new, on MyModule. And when, the syntax is exactly the same for calling the method, because ALPtr class implements the operator arrows. So, it was exactly the same as python. OK, just little remark about the syntax. So, now let's go back to what I was saying - Specific proxies. So, we also provide specific proxy. This is, for example, specific proxy on motion module, on ALMotion. We have ALMotionProxy instead using direct smart pointer of it ALMotion. We writing New on this proxy with the name of I'm sorry, we just add getParentBroker.

4.7. Module development

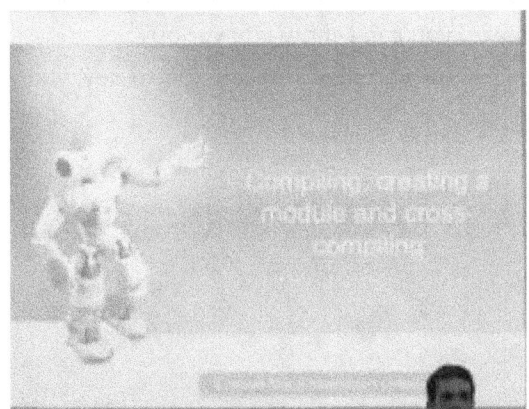

And then, once you have that, you can, actually, call your walk method directly, instead of using call method So Basically, and have ALProxy here, which has four methods. Method name, parameters. But we also have it's because the method ALProxy and we have ALProxy. And this class implement all the methods of the all different [?] methods of the module.

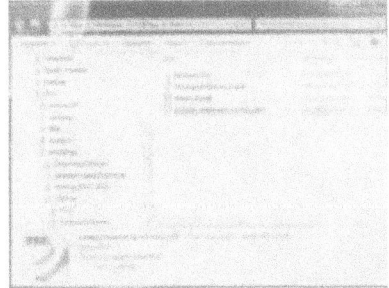

So, here, for example, here you have method. OK. So, the difference will be in if you use generic proxy and you want to call a method Walk of animation [?] from here, like that. Proxy pointer instance Call, and then first parameter is the name of the method Here method is Walk. And if you have a specific proxy, because the walk method is implemented, you can directly call Walk. So, of course, this one is better, because here, for example, if you let a little mistake and write instead of it will compile, but when you execute - will not work. That's why there is written "less chance of typos". And also quicker and it enforce the of the parameters, types of the parameters at the compilation time. So, now we should all specific proxies for all the modules we provide. But if you write your own modules, of course you don't have specific proxy. So, you can use generic proxy, or you can re-implement specific one. To also use specific one. And that's it for now, I will show you how to do your module, compile it. Do you have any questions so far about NAOqi?   So, we provide some example, some different module example in documentation, but we only provide the source code and we don't provide runtime files, because we want to stay cross-platform. So, you want to compile it, you can compile it on Windows, compile on Mac, on Linux.

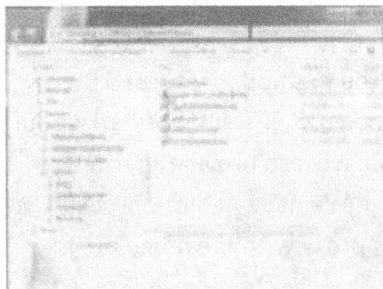

So, if you want to compile it, you can use a tool which is called cmake. Windows cmake. Yes so, cmake, you can download cmake on a "cmake.org» here external software installer of cmake. So, the first thing you need to do is to install cmake. Once you have installed cmake, you can I show you various documentation. You also, of course, also need to install SDK. I'll show you this. You can do usb-key, you have this folder, right, external software.

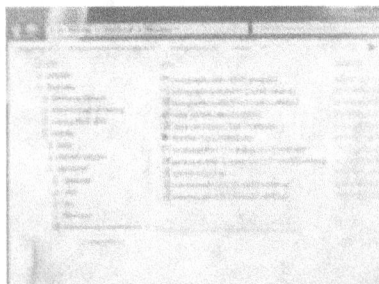

I've put some software that you can download on the internet.

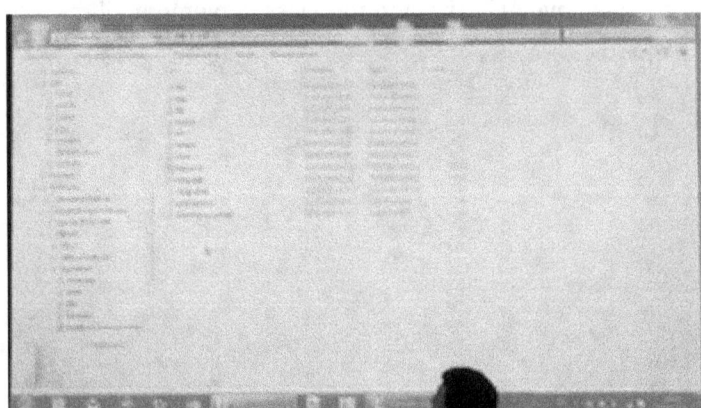

And, like I said install simulator there is C and python and in NAO install, depending on which platform you are, we are on the Windows, we also have the SDK. This one.

This file, this called NAOqi SDK sorry, this SDK is for and, in fact, you have also SDK for Mac, and also SDK for Linux so, depending on platform, you just unzip the SDK when you unzip the SDK, looks like that. You have these files. Basically, you have, on Windows, you have, but I think it's similar on the other platforms.

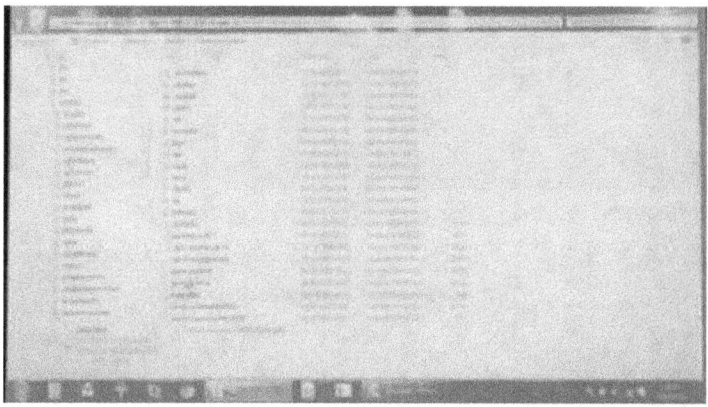

On Windows you have two batch files to launch NAOqi, on Windows. To launch simulated NAOqi. If it launched you have NAOqi running, it's just program for windows, not very exciting just loading of the modules. You have folder with some binaries, here, executable and so on. You have a doc folder, the include or the header files you have all the files for specific proxies, like I told you.

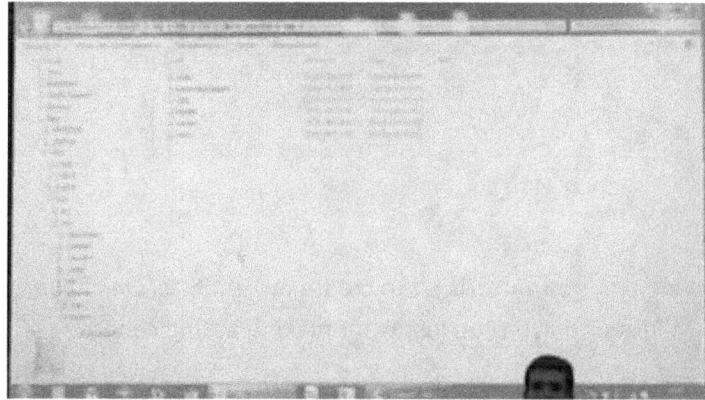

Library folders, where you have all the libraries compile all the modules. And some other stuff, not very important. So, in the doc folder, this is where you have examples. In the doc folder you can launch the is the same documentation, as that I show you yesterday. And if you go to examples, when you can find some examples. In C++ or python. So, let's look into C++ examples. It subdivided into category. You have example for different modules. or Animation modules, vision modules, sensors. And you can look into core

In core modules we have, in core you have hello world or say hello world If we look into module. In module you have myfirstlocalmodule or myfirstremotemodule. OK, so two examples to create your module. Local module. So, here, we'll take myfirstremotemodule. It's folder that should not exist. Like I told you, like I told you just before, we only provide C++ files, source files, headers and C++ files.

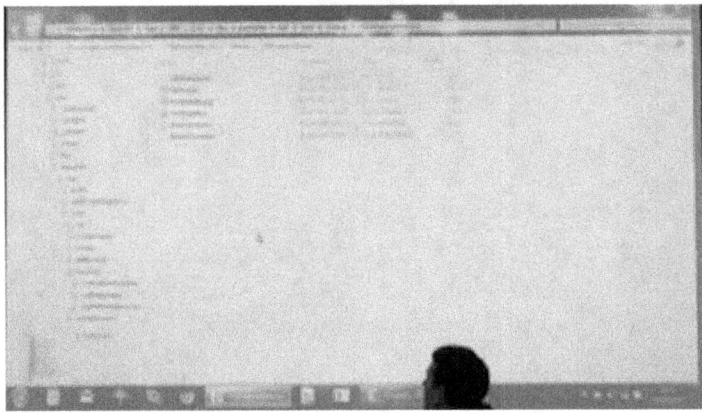

While we also provide some files for cmake. File for cmake, because, like I said, we use cmake to build project files.

So, it depends on which platform you are. In windows, you wizards [?] to create project files. On, if you are on Linux, you create, maybe, solution for Eclipse or maybe you just want make files and choose cmake. And same thing on Mac.

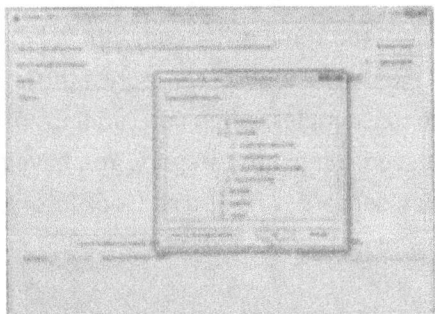

So, I already install cmake on my computer, so we just run it and we started it, when we start cmake it looks like that. So, it's very simple, just need to specify where is the source code for which you want to build the project SDK doc example c++ core module myfirstremotemodule.

I will take the remote one I can also show you OK, and when, this is there the source code are. And when, you should also specify folder, where you want to generate project file. You cannot use the same folder.

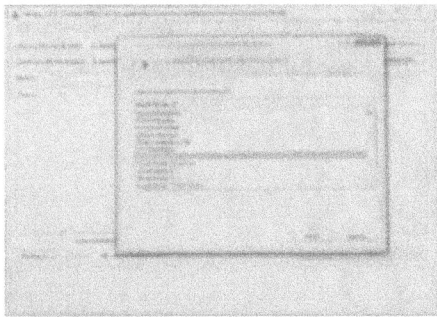

It's a limitation for cmake; otherwise it will exit showing error. But you can choose subfolder and once you have done that, you can click Configure [?] button.

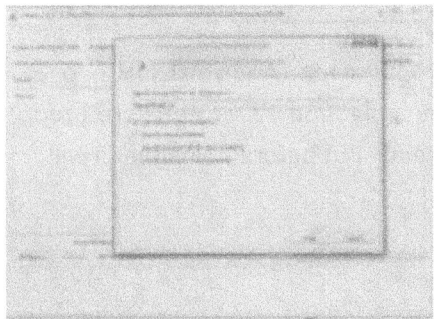

Cmake mistake says that folder doesn't exist why don't create it. Create this folder. And then, you should specify for which on what IDE you will run your project. For example, Visual Studio 10 or 8 or whatever, or Eclipse, but, we don't support all the compilers. When you download the SDK we don't support Windows 98 and the same for compilers So, here, of course you just choose Visual Studio 10[?] And then, you should also choose this one, so, why this one? Because cmake tries to create a project file but if you do that,   that's just do   compile We have, we provide, at this day we provide, configuration files for cmake.

Our engineering friends [?] have provide configuration files for cmake. name of your project, and which settings you should place in a solution, in a project. So, you should specify project file, or project files and click next, then you should specify cmake file.

Cmake file is located at the root level of the SDK. Here, on my computer it's called   I exit my SDK here, if you remember then you can click Finish. And cmake is generating some files, here. For Chinese users. Do not unzip SDK into folder, which contains Chinese characters.

Because cmake does not support Chinese characters. Just use English characters. And you can so, once it have finished to emit some stuff [?] you have a lots of red stuff appearing. And it's written here, red generated so, once you have red stuff, you just press again, again, again. But before doing that, I will also show you just press configure again.

And now everything is white generates. And the generation is stopped. So, now if I look into my folder, I can see, that cmake has generated some project files and also there is solution for Visual Studio.

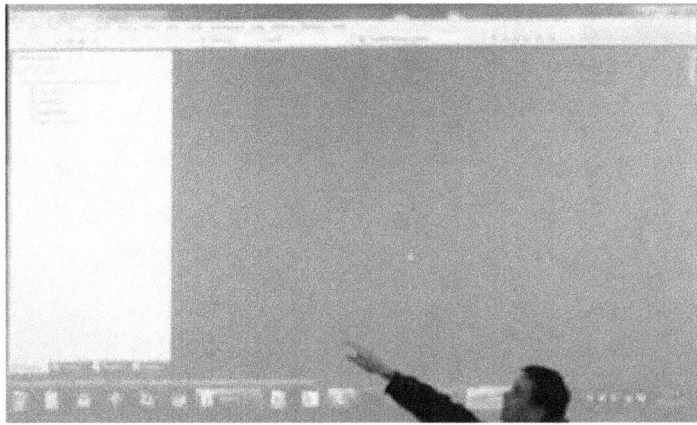

Here, I can just double-click on the solution   First thing you need to do, is to set your myBroker solution, oh, myBroker project settings as Startup project. Cmake cannot do that.

In Visual Studio, the Startup project is a user defined settings. So, you choose Set as Startup Project, otherwise you will not get it. OK, and when it appears in bold characters.   So I can try to compile. Just press F7 to compile.

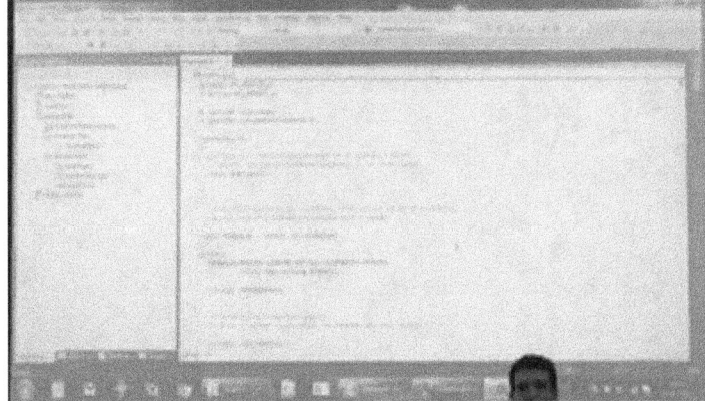

And   to compile it   Lets come and look at the code, if I compile the Broker project. Header file, source files. Install the header files. So, this this is project, just created module,

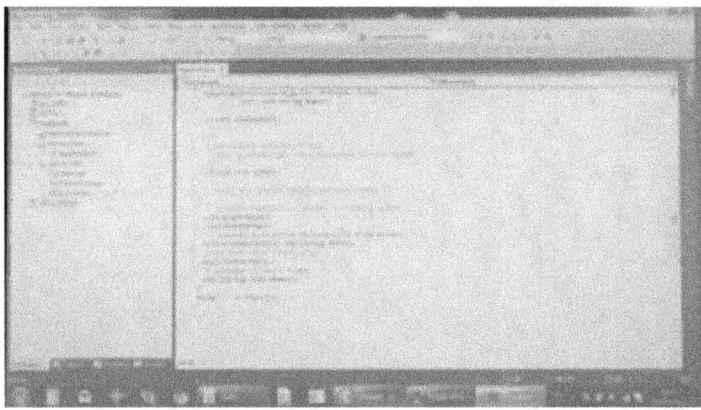

Which name is MyModule? All that which inherited [?] from MyModule OK, in that class I have the constructor, the destructor, and then some methods, like test method Now, if I look into C++ file, for the constructor, I can see exactly the code I shown you in my application in So, we call function Name, function Name is a method of a base class AL [?]

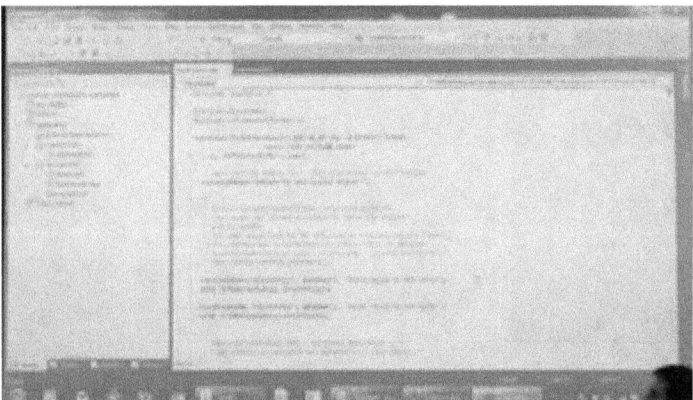

Calling that from base class. Function name, arrow [?], name of the module. Actually, there is function called get name, and it just return the name of the module, you can probably use it, it's safer. And then the description of the method what is important?

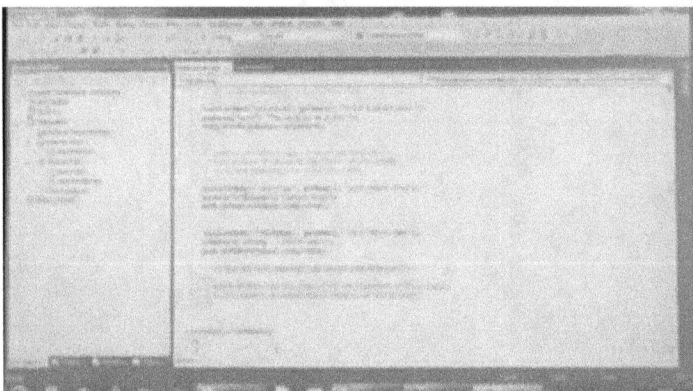

There is my method and arrows. You can do the same for another function - test method, exactly the same. On top of constructor you can also see setModuleDescription, this method is also method of base class and this one is to give documentation for whole module, general documentation of module that would appear on top of documentation.

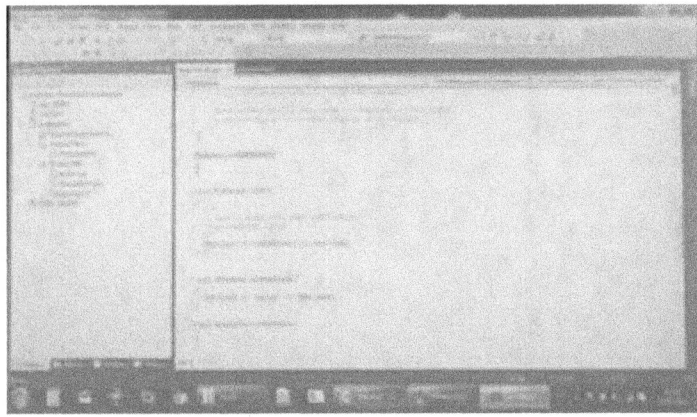

And, this function takes the parameter from and this function return value. And that's it, your constructor is a functional module. Only binding the methods. The destructor - nothing.

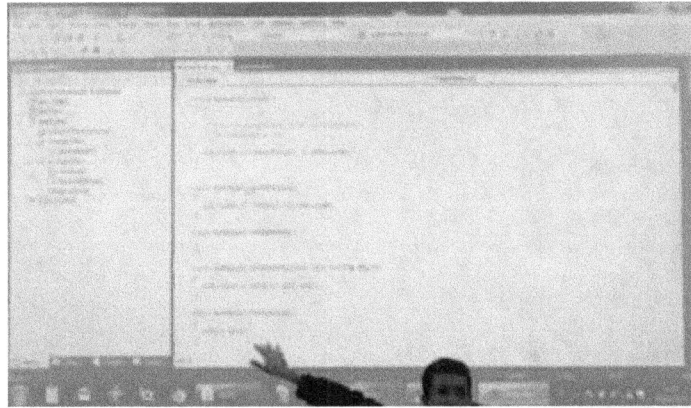

Initiation function - just method count. And then, we can check the method, the method itself. But, since it's a very basic example, almost empty, we can just make count "Hello", just print hello in STD output.

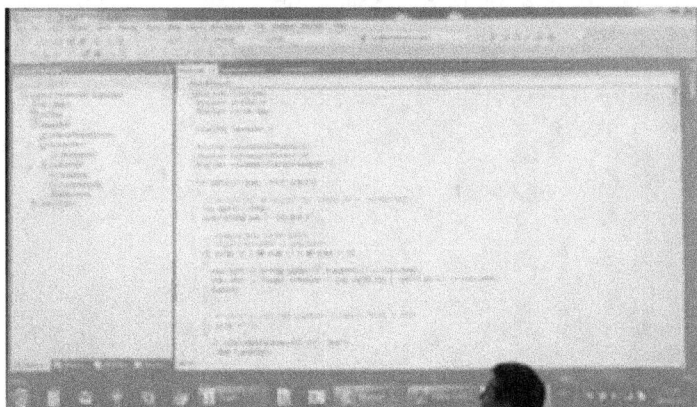

This method just doing nothing. Taking the word parameter and just display the word. And the return name, just return true. We can have a look into the main C++ file, if you are curious about that.

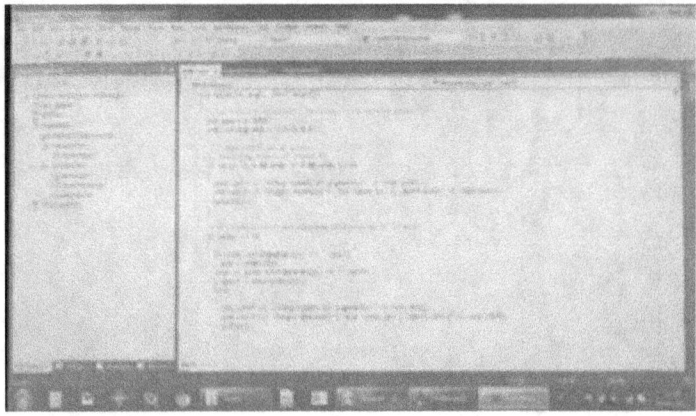

So, in the main you have a method which takes argument why? Because it's specific IP address of the robot you want to connect.

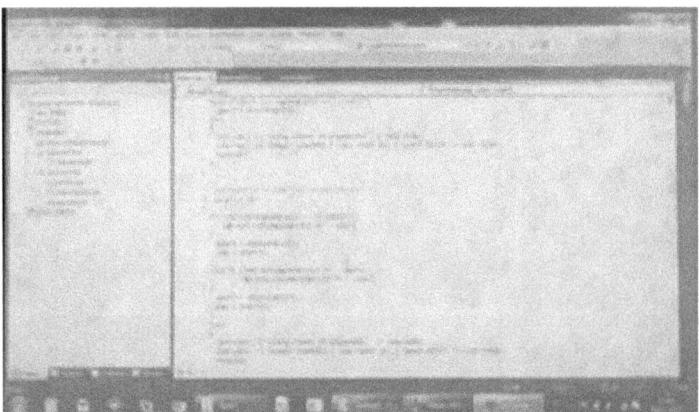

So, here, for example, we have a port and address, ip address, which by default, default port and by default is a local host. And we check arguments is either nothing or is a and we check, if we have argument --pip, in that case we just call --port, in that case we just get ip address and port. And otherwise for here we nothing except just checking arguments of the application. Then, finally, I told you about remote module

on windows broker. You see there is in front of create Broker. You just need to give it a name and then you just call function create Broker create and store it in a broker variable, which is a smart pointer on an ALBroker and the shared pointer - very similar.

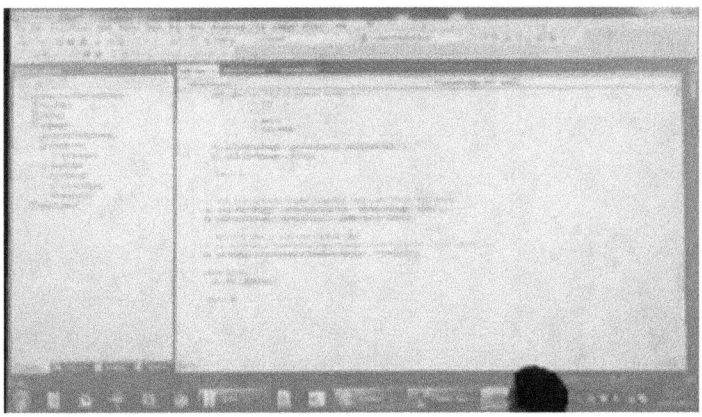

So, it create broker. Catch and then, we can add broker to the broker manager and, finally, we create a module and attach it to the broker we just create. And give it a name MyModule.

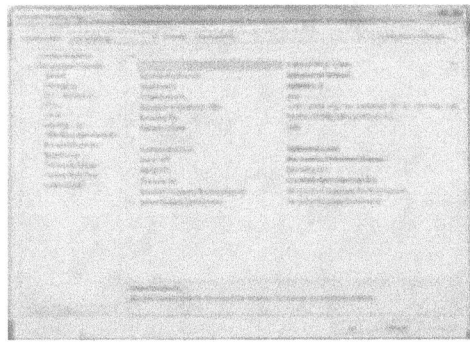

So, that's it, basically, this is what's the while, oh sorry, the main and, of course, we just do exit to the program and enter into loop [?] and so we kill [?] the program. So, I will try to connect to that robot «Hello, I'm robot, my internet address is 192.168.1.103, my battery is fully charged".

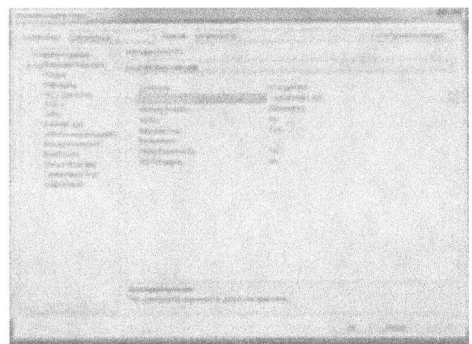

So, I will close the project. And in the debugging I can specify the arguments, common arguments.

So, here, I will just give ip address. And then, I will try to run the robot. Now, we have MyModule running on my computer and connected to this robot.

So, you can check it, if I go to NAO page NAOqi that my Module is connected to my robot and if I click on it, I have documentation and I have documentation for module.

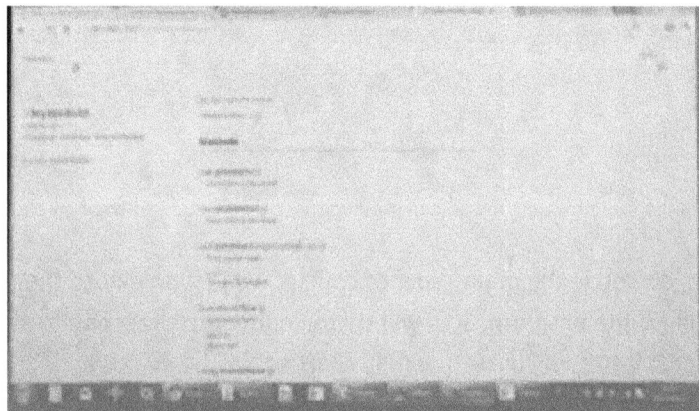

This text here, you can read it exactly same text as what you have here. Load this text you have, you have here is documentation about all the functions in your module.

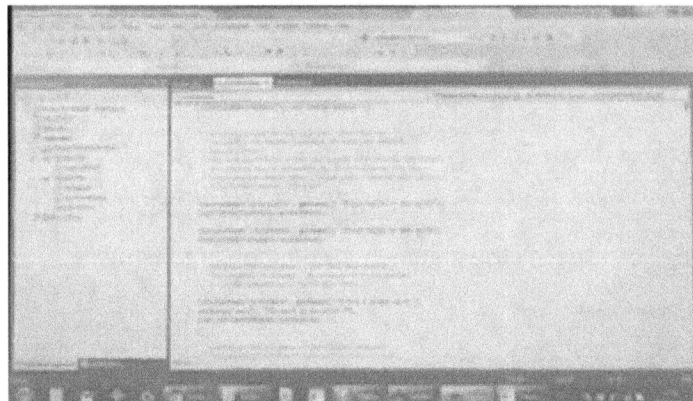

Its running [?], let's it off. So, now we can try to use its methods, we can try to call this method. Just remember - module, is just a library. Is library, it just provides him services. Then later you probably want to program, using those services.

OK, so I can do the program in Choregraphe. I will connect to a robot here.

And I can try to call one method for my module. To do that, I will do a simple python script box.

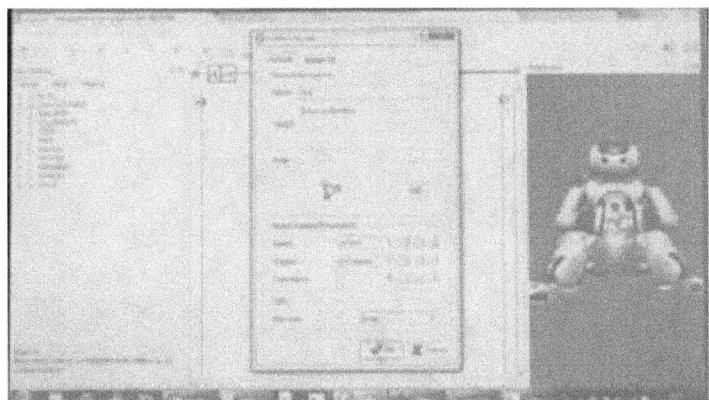

OK, that name box. Type which is script, ok, script. I will edit the script. In the on Load I will create a variable; I get a proxy on the module. Proxy. I should give module name, so module name is MyModule capital M, just copy-paste

Ok, I told you, that the other parameters are optional, if you don't specify it's in your own program[?], you asking it's on the robot where this program is running.

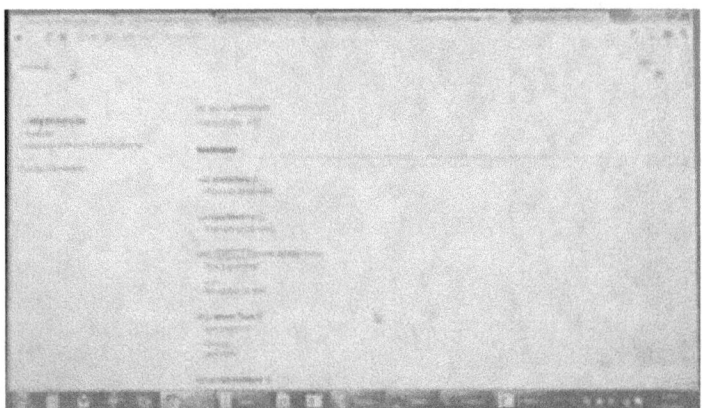

Ok, I will store, actually, this proxy into the self.

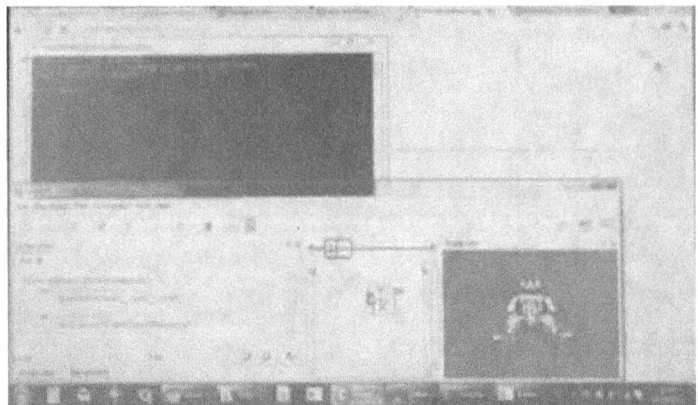

Self-proxy why I store it into the self? Because I want to use it. But, on Start input. Then, you can call the function.

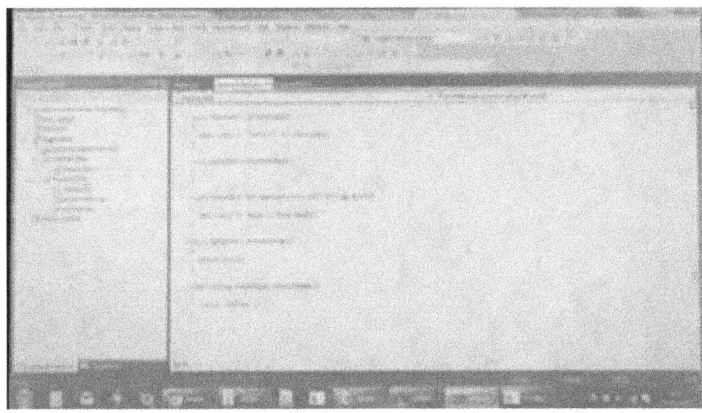

For example, I can call   print Word. OK, so, now I can run my program on the robot, OK, it's running, I didn't link the input, because I just wanted to show you, currently there is nothing here. And now if I trigger the input of the test box, then it displays the words   in the stored output. Of course, your module is running on your PC and it's running under Visual Studio, which means, you can also add breakpoints into this function. I can add breakpoint here and trigger the input again. Then there is Visual Studio break, so you can debug. You can simply like this, value of word. OK, any questions? It's clear, everything is clear? [Quiet question from listener]   Sure. You can do it, but this afternoon, you can just reserve this question to practice. You can   and it will answer your question. How .if you are going to group, together?

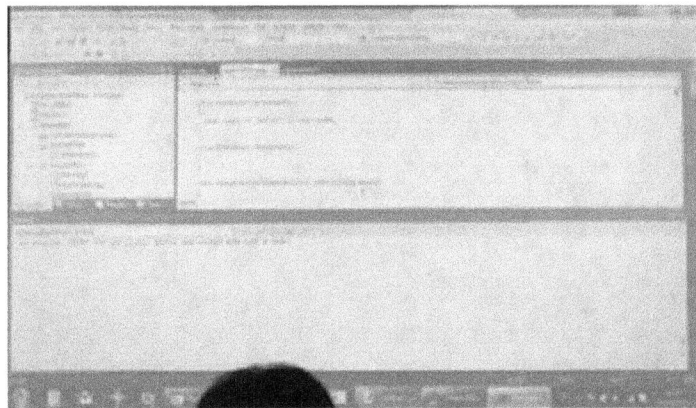

Think   here? What about folders? Create groups? I think, it would be nice, if you can divide in groups for this afternoon.

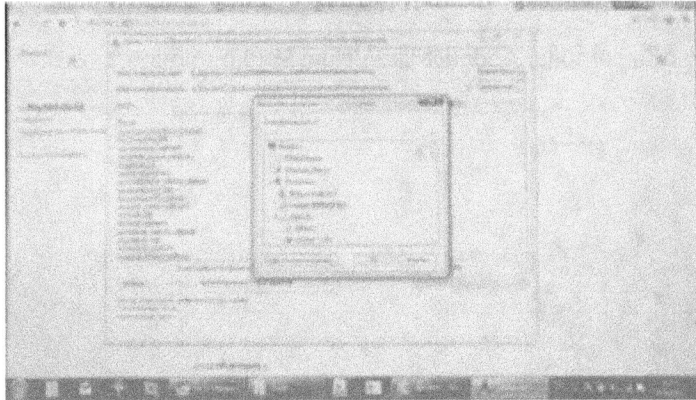

But you can start so I can go a little further. I have say, I show you remote module running on my PC. I can also show you the local module running on my PC. That's maybe

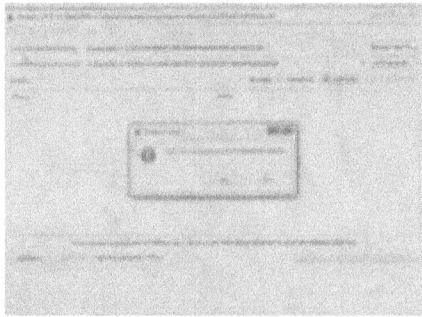

So, in cmake I will also create foundation [?] for local module. This time myfirstlocalmodule. And specify local file Then I click Configure. So, we do it several [?] times for you, to remember Yes, I will create the build folder, and choose Visual Studio 10, specify local chain. Chain file, option file is here, SDK.

First creation [?] Again Press generates. Generation is done. Let's go to first module. Here I open the solution. I choose my module as the startup project. I can try to compile now. And, oh I'm sorry, I pressed F5 instead F7.

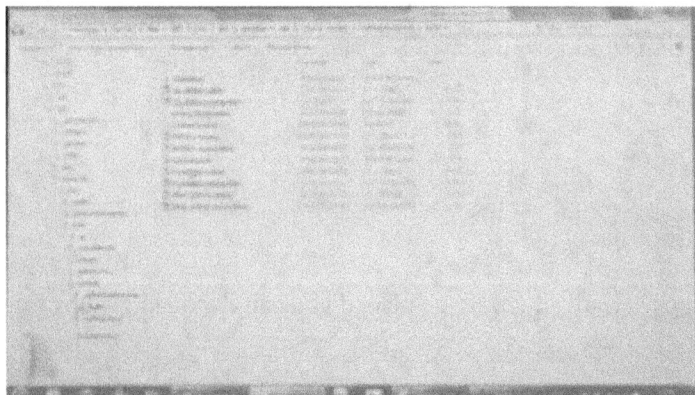

Started debug instead of compile. And I succeed to compile. And just like Visual Studio told me, this time we generated DLL. Library generate DLL file.

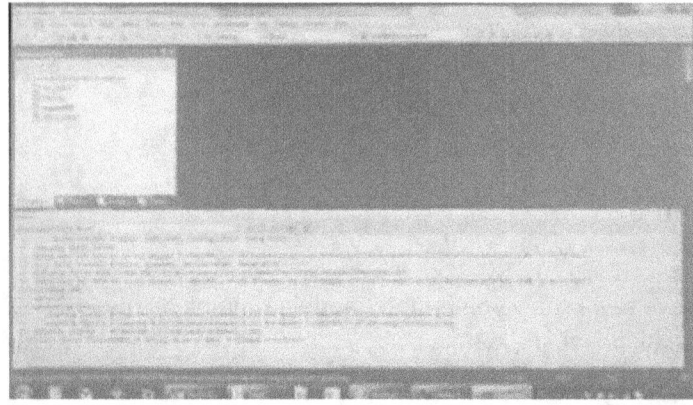

We can look into the header file. We will see, that it's very similar   MyModule, of AL Module Same kind of methods that would bind The C++ class is exactly the same.

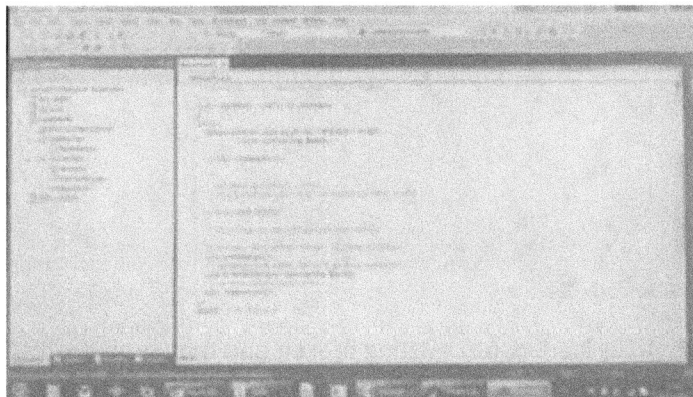

Surely, local or remote call doesn't change, it's only   is broker, there is broker on my computer [?].   And the function is the same. So, this two are the same. Only this change, so, this time we are creating DLL. For familiar of compiling DLL.

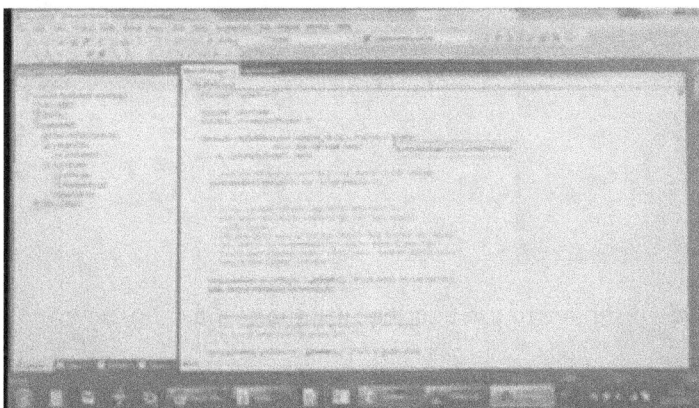

But you should use this kind of syntax. This declspec dllexport. And you have to specify create your module as one entry function of your DLL. And if this is functional, the DLL. We can find again the creation of the module.

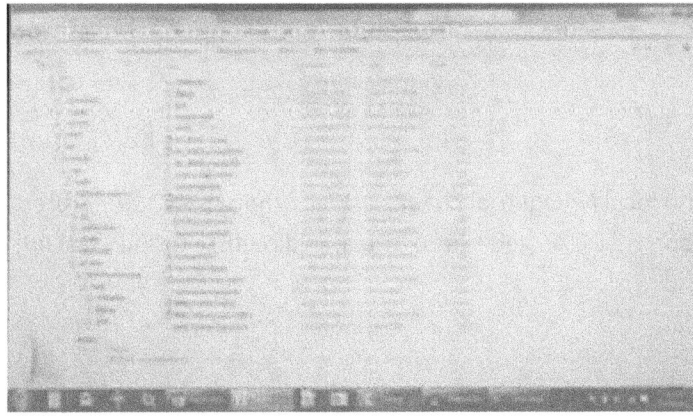

But this library doesn't create the broker.

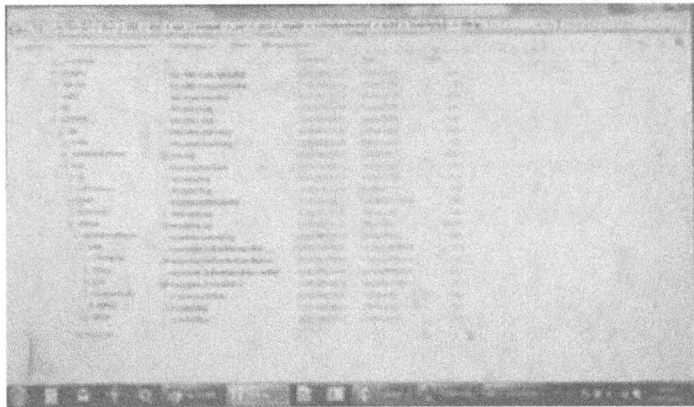

It just get broker as the parameter. Because the DLL is loaded into existing broker, and then broker will call this function there   broker is here.

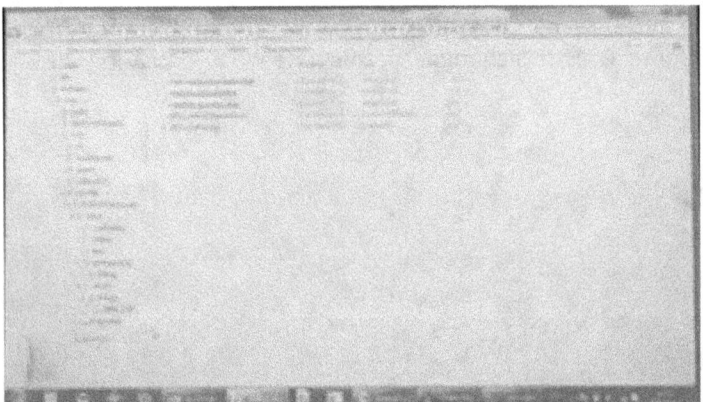

But it's very similar, similar code. That's it, so you just have to use declspec instead of main. So, now, how to load this. I'm searching

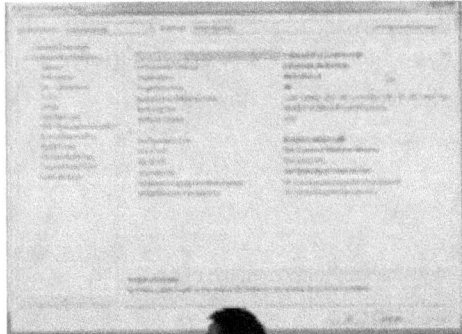

Oh, yeah I think it's generated differently for the local NAOqi tool chain [?] Yes, otherwise it's generated here. Don't have changed it in one parameter so; the DLL is generated into a, in directory to the   NAOqi folder. Like we have   so we can change it.

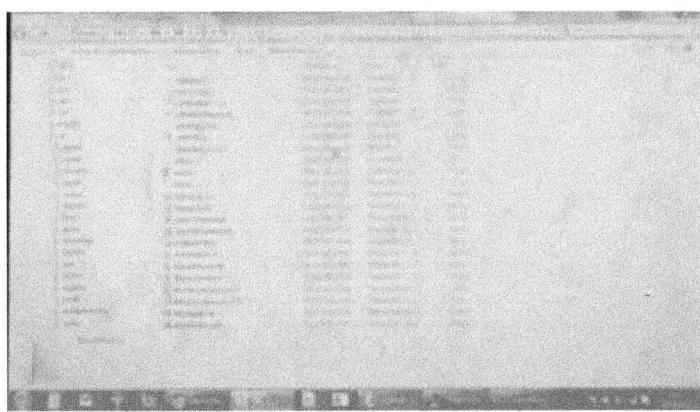

So my module is here and now, if you want to have this module loaded into the local NAOqi, you have to specify it into ini file, which is named afterward   and it should be   here

I've lost it on my PC Ah, it's here.

So about other configs, I'm not sure it's here so, here you should add the name of your module that you want to load and launch.

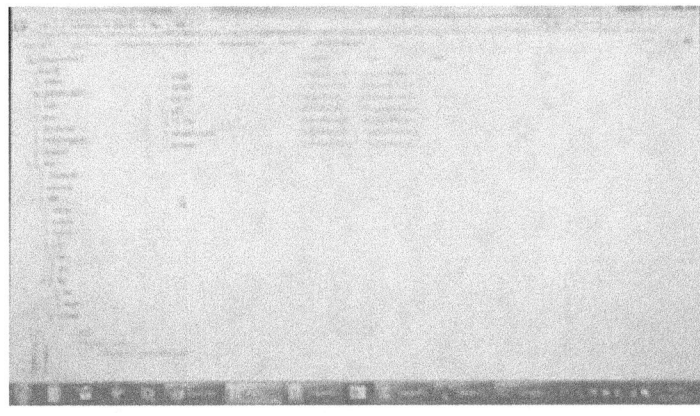

I will try, not sure, it's like that. Before, it was like that. So, your module called is MyModule and now I'll try to launch it. Here, OK.

So, its loaded module MyModule is started. So, I'm not fear [?] Exactly such method. To summarize what I done. Is, here, I compiled my module, my local module. So, local module I compiled for PC, for Windows.

When you compile local module it generate compiled, it generate DLL - dynamic library.

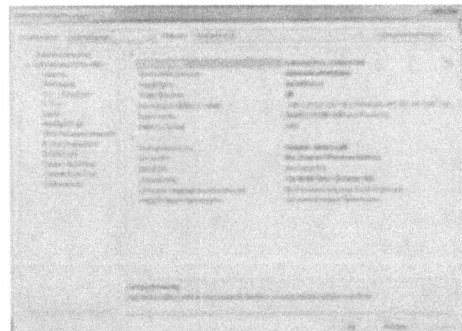

You can see that this dynamic library is generated directly into the lib NAOqi folder of the SDK, where you have unzipped SDK. I have, what I just compile my code, I have generate my DLL and it's placed there, in this lib. Then later, I need to tell my local NAOqi, my simulated NAOqi, I need to tell how load my module, my DLL into the NAOqi. So, for doing that, you need to modify ini file, which is called auto load, "autoload.ini" and located in the etc., etc. NAOqi folder of the SDK.

I just give the name of the DLL I want to load, without the dll extension. And then, when I launch it, I just launch my local NAOqi.

It loaded all different modules and also my module. [Quiet question from listener]   Debug it   NAOqi understand this local library NAOqi So, now I can test it.

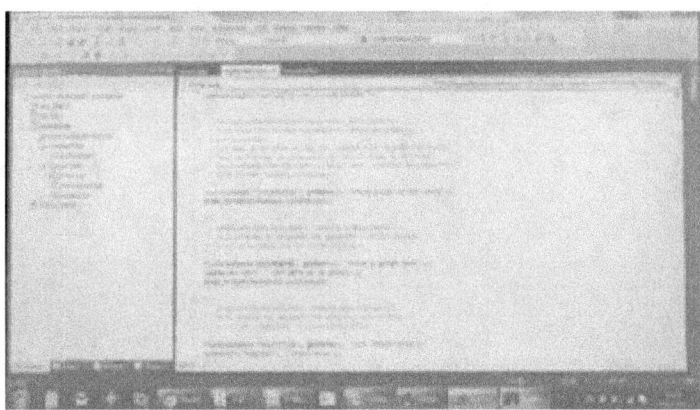

I will test with the same program, I think, I think possible still exist NAOqi yes it's still there   MyModule and   load method

So, this time, I connect to the local box [?], which is the NAOqi running. Choregraphe is just connected to it, here. And you can yourself see some of it above edit and then I can launch it, and then I can try to trigger my input, but, I think this time, in output window of NAOqi, because it's a DLL library this time.

Basically, that's different. Any questions? Now let's go a bit further. Actually, you have a robot now. Once you have connection, you have loaded your modules with some great functionality inside. Its working you want to put it NAO environment. To do that, you can, first you need to decide if you want to do remote module or local module in NAO. OK, at first, I suggest you use local [?] module, it's faster. And that means what? That means, you need to compile your code for Linux and to do that, you need to compile under Linux. So you should install Ubuntu on your computer, this is Linux. Or you can use virtual Box. So install Ubuntu, or take it to the Ubuntu. You copy your code to Ubuntu, you use cmake for Linux, OK, so you need cmake for Linux, install cmake Launch cmake, generate project files for Linux. Basically, it generates some make files. OK, just make files. And you compile it for Linux. And instead you get an SO file. On the Windows, if you build local module, you will have "myModule.dll". If you build remote module, you will have "myModule.exe", exe file. On the Linux, if you build local module, you will have "myModule.so". That's the extension for the library. If you build remote module, you will have "MyModule", no extension. Files on Linux don't have extensions.

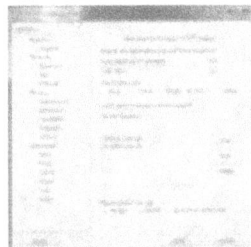

So, basically, you take your Ubuntu on your computer, OK, Linux Ubuntu. You get your source code, cmake under the Linux.

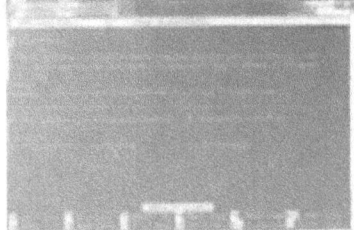

You build your project file for Linux, then compile under Linux, then you get SO file. Then you can take this binary SO file and copy it, transfer it to NAO's head you can put it wherever you want, in the home NAO, for example and then, once it is here, you also have auto load .ini on the robot.

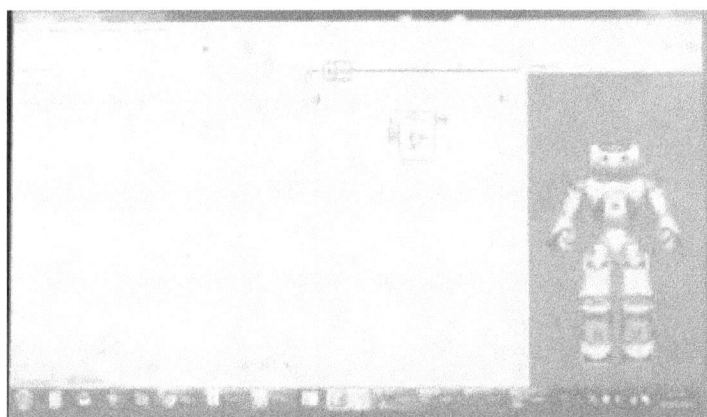

I will show you, how to do this. [Quiet question from listener] Local on windows I'm, yeah Local on windows it's for simulating NAOqi, because binary for Windows, but remote under Windows is for Windows it can only work under Windows Well, I think it's inside and here you have all So, that is quite similar, a bit different, but

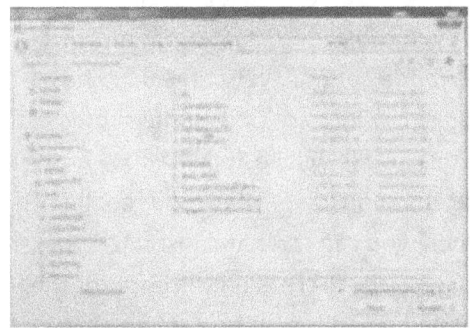

So, you can, a user, you can type the path of the SO files library.so so, you can, if your SO file in the NAO home, home NAOqi just put it here, and then once you robot reboots it will launch

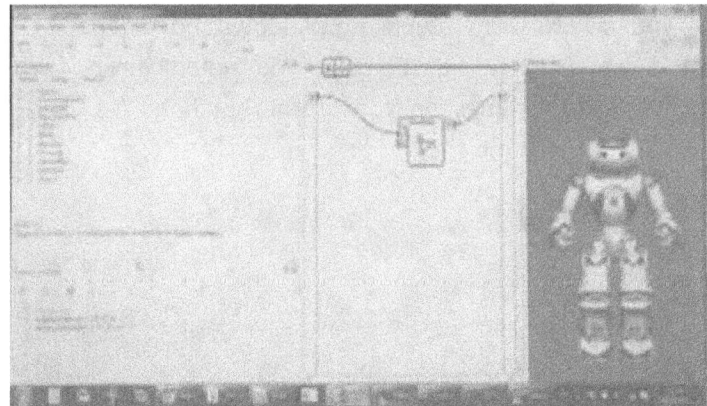

That's start the main and launch but you can also do it on the fly, by your program. So, I will show you something like this. Start my program, I will open project OK, so I want to show you this. If you remember, yesterday I had talk about project content and example I give you is music files, but in this behavior, we actually put SO file.

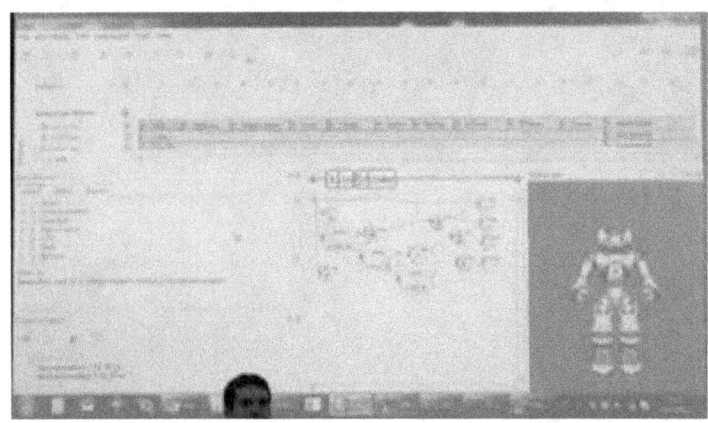

We actually put there library, that we have built for Linux. We put it inside content. Then, in these boxes, you have some python script to load, so where, I don't know where exactly loaded this, but you can imagine you have python scripts somewhere, who loads this library. Who load this module? Cause module can be loaded on the fly. So, if you don't want to, after creating your SO file, if you don't want to copy it on the robot, and modify control that loading and stopping it - you can also just put it inside your program and make the script box to load it, and if you do that, it's even considered better, it's easier to debug. Because, if you copy it on the robot, every time you make a modification, every time you find a bug, you have to again copy it, and you have to restart NAOqi. Restart NAOqi to reload module   but here, if you just copy in that folder, when you press replay button this SO file will be "pewit" downloaded into program and run it, unstopped. So, it's easier, because you just, even if you rebuild that, you just need to   any questions? It's clear for everybody? When, you can try it and just unzip the SDK and try it. For the SDK there is no license issue, so no warranty on SDK   there is only license for the Choregraphe program SDK is just a zip file, so you can just unzip it and if you are   first start you can just compile it, it will works. [Quiet question from listener]   Choregraphe suite includes? Yeah. Choregraphe suite includes Choregraphe and not the simulator. [Quiet question from listener]   the simulator is? No, there is no NAOsim for the Because NAOsim is   NAO compiler, since   [quiet question from listener]?   Probably it will work. Because, you have NAOsim running on Windows and   on the Linux but you have to specify another localhost any questions? Then, actually   you should launch NAOqi and start C++ programming and you have time   [quiet question from listener] python? Yes. Oh, yes, python programming. So, in python you can also write some

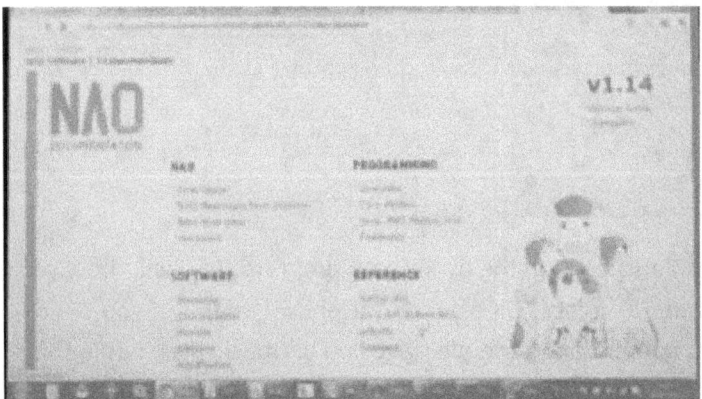

you can write some module in python.   Python module so, you just want to make python code, you can just use a python editor. And then you can start to write some code, like that. Basically, you create a class which inherited from ALModule. Possible, base class is but compiles it. And, after that, you have initiate And, after that, actually, all function here, all the methods here   for example we can define also the main, if you want to make module   In this script we define   in module, and also main to run this script as a and if you just want to make module on the module if you look [quiet question from listener] if you create, if you use python script inside box? When I would say, it's more python script for your application our module so, you can do both. You can also, again, of the same logic of the file in Choregraphe, which means we can have here, you can, in your project content, you can include the python script here. So, we can have python script here. And with some very simple script box, you can execute your python script, which can be just normal script, or which can be module. So, if it's a module implement in module if it's just some script

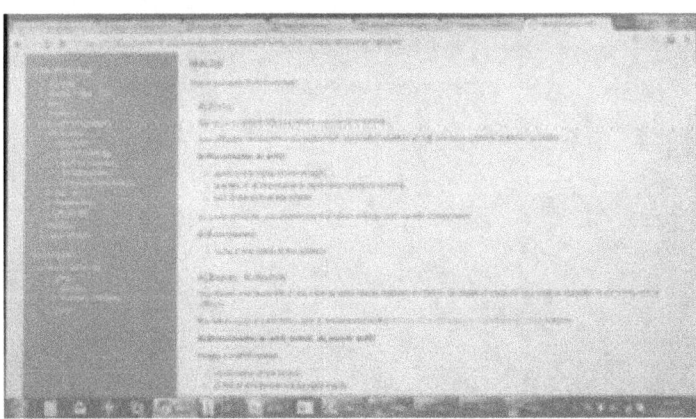

I will show you the API.

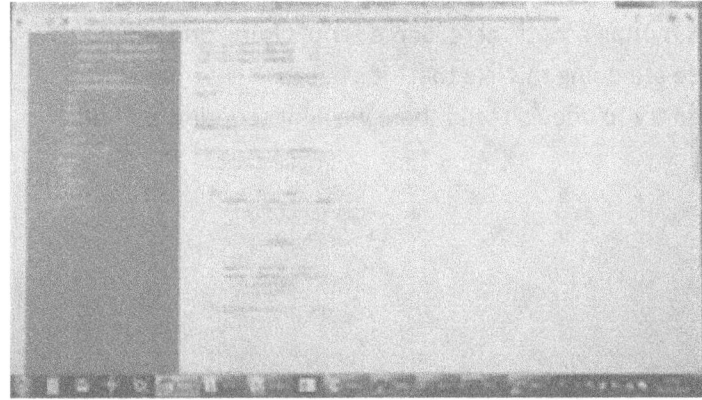

We have a so, this tutorial I just show you how to do and the NAOqi API have modules like ALLauncher API

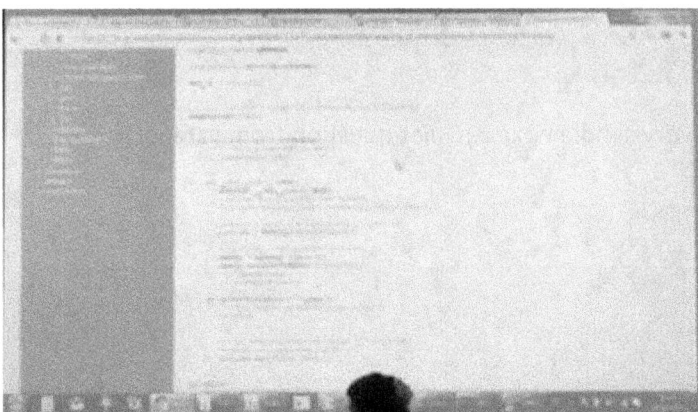

you can launch, for example, your python module here, and you use the name of the module. I'll show you I'm sorry, the method because in 1.10,

 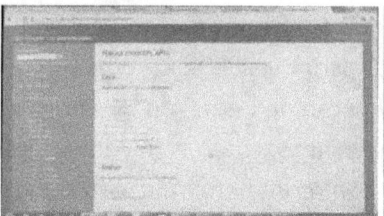 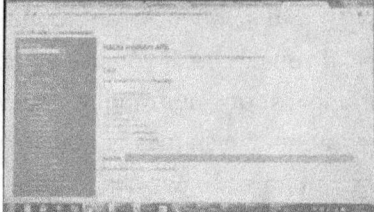

because we use ALLauncher. And NAO is also

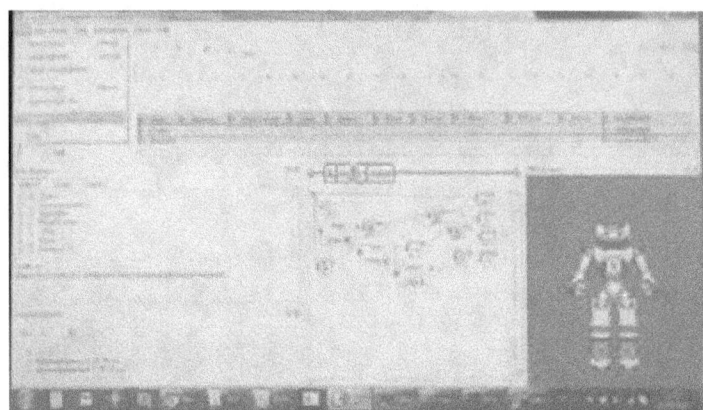

[Quiet question from listener]   And I know, that some people prefer to do that what I mean, I show you example.   As you can see in a, here for example, I just opened mouse library [?] As you can see in project content we have some scripts, which is python scripts, here. Some scripts here, and when you have box to load and execute them. So, why we are doing this like that? It's because   user enter external   so it's smaller, it's easier, we can even try to debug it and then, we limit scripting, python scripting.

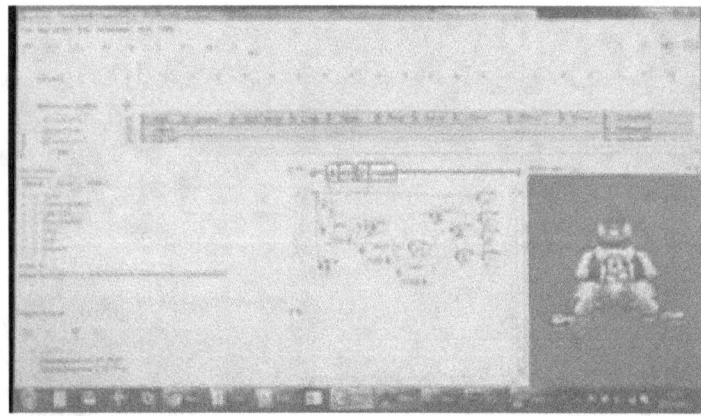

Choregraphe introduce what is necessary for very simple boxes. [Quiet question from listener] behavior scripting dates?

Dates, well Yes. How do you launch this behavior Yes? So, basically, yes, some people like python to do like that try to see Choregraphe as a very high level programming tool. So, it just makes your program very simple program with some images, like that. And all your algorithms create either in C++ or in Python.

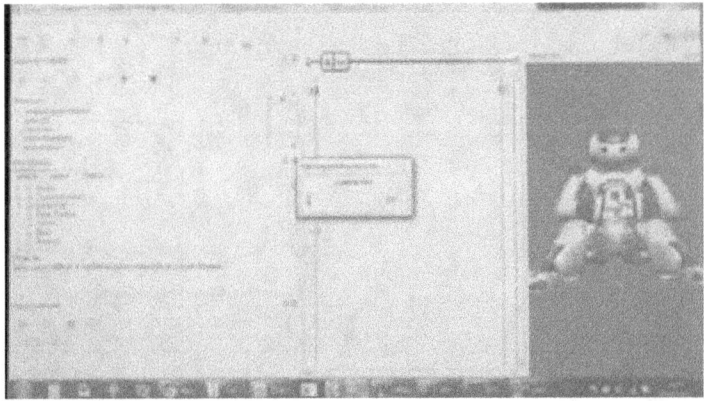

For example, if you want vision algorithms fast you should do it in C++ services and then you can use high level [quiet question from listener] Yeah, sure. Any more questions? So, again, this afternoon it's all about questions, your questions and maybe practice a little bit. Installing the SDK, trying to build something and that you start, any groups.

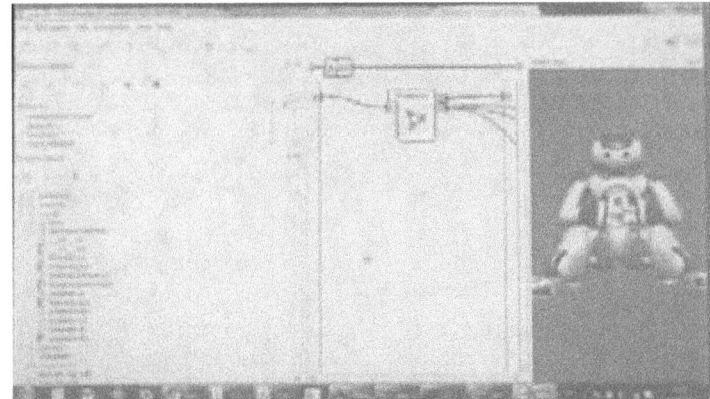

So, just one group here, one here. Try to make your groups, and try to start your project Because, if you have the group and you start, you will ask me, you can ask me - how to do this, or what But, for that, you need to have the group set, you need to have start discuss. So, this afternoon would I like you to have any groups in any space. Each group take one of table and start discuss. Every time you have question, rise hands. OK? So, thank you. I'll see you at half past one, is this correct? [Quiet question from listener] Just I just said the group